CAMBRIDGE STUDIES
IN MATHEMATICAL BIOLOGY:
Editors
C. CANNINGS
Department of Probability and Statistics, University of Sheffield,
Sheffield, U.K.
F. C. HOPPENSTEADT
College of Natural Sciences, Michigan State University,
East Lansing, Michigan, U.S.A.
L. A. SEGEL
Weizmann Institute of Science, Rehovot, Israel

MATHEMATICAL ECOLOGY OF PLANT SPECIES COMPETITION

ANTHONY G. PAKES and ROSS A. MALLER
Department of Mathematics, University of Western Australia, Nedlands, W.A. 6009.

Mathematical ecology of plant species competition: a class of deterministic models for binary mixtures of plant genotypes

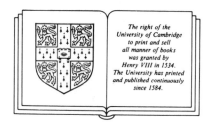

The right of the
University of Cambridge
to print and sell
all manner of books
was granted by
Henry VIII in 1534.
The University has printed
and published continuously
since 1584.

CAMBRIDGE UNIVERSITY PRESS
Cambridge
New York   Port Chester
Melbourne   Sydney

Published by the Press Syndicate of the University of Cambridge
The Pitt Building, Trumpington Street, Cambridge CB2 1RP
40 West 20th Street, New York, NY 10011, USA
10 Stamford Road, Oakleigh, Melbourne 3166, Australia

First published 1990

Printed in the United States of America

ISBN 0-521-37388-3 hard covers

British Library Cataloguing in Publication Data
Pakes, Anthony G.
    Mathematical ecology of plant species competition.
    1. Plants. Ecology. Research. Statistical methods
    I. Title   II. Maller, R. A.
    581.5′072

Dedicated to
Dr R. C. Rossiter

CONTENTS

Preface   ix

1   **Introduction**   1
1.1   Introduction   1
1.2   General description of the models   4

2   **Mathematical formulation of the models**   11
2.1   Introduction   12
2.2   Formulation of Model 1   12
2.3   Formulation of Model 1A   14
2.4   Formulation of Model 2   16
2.5   Constant soft seed probabilities: approximation to
        Model 2   20

3   **Analysis of Model G: Introduction**   25
3.1   Introduction   25
3.2   Rudiments of dynamical system theory   28
3.3   Some preliminary results for Model G   30
3.4   Review of related work   34

4   **Analysis of Model G – no interior equilibrium**   53
4.1   Introduction   53
4.2   Global asymptotic stability   55
4.3   Determination of $\mathscr{C}$ for Model 1A   61
4.4   Convergence rates   69
4.5   The case $\rho = 0$ 77

5   **Analysis of Model G – one interior equilibrium**   79
5.1   Introduction   79
5.2   Stable interior equilibrium   81
5.3   A Liapunov function for Model 1A   86
5.4   Convergence rates   87

5.5	Unstable interior equilibrium	96
5.6	Coincident isoclines	101
6	**Analysis of Model 2**	106
6.1	Introduction	106
6.2	Some preliminary results	108
6.3	Attraction of equilibria when $cd = 1$	111
6.4	Long-term behaviour when $cd \neq 1$, but $N < \infty$	120
6.5	Local analysis of the infinite memory case	126
7	**Application of the models**	131
7.1	Introduction	131
7.2	The standard data	132
7.3	Applications of the models	133
7.4	Dominance change in Model 1A with parameter variation	148
7.5	Comparison of models and parameter variations	151
8	**A single strain model with cropping and tillage**	158
8.1	Introduction	158
8.2	General description and notation	160
8.3	Mathematical formulation	162
8.4	Asymptotic analysis	172
8.5	Application: the effect of softening rates	180
	Bibliography	186
	Index	189
	Index of Notation	192

PREFACE

This book is a graduate/research level case study of the mathematical modeling, as opposed to computer modeling, of the dynamics of two competing plant species or strains, or genotypes. The process modeled in fact is a binary mixture of two strains of the annual legume subterranean clover (*Trifolium subterraneum L.*), which propagate by means of "soft" or germinable seeds each year, whilst retaining a store of "hard" dormant seeds for germination in future years. This process, using hard and/or dormant seeds, is common to many annual plants. The interaction between the two types is described by the de Wit replacement model.

Our models were developed in response to four major questions on long term outcomes of binary mixtures: (i), is ultimate coexistence possible?; (ii) if not, which strain will win?; (iii) does the mixture approach an equilibrium?; (iv) if so, then how long does the mixture take to attain it? These questions were motivated by the particular case of competing strains of subterranean clover which, although of Mediterranean origin, is now widespread in southern Australia and in some parts of the USA, and is of great agronomic value and interest.

If two strains of this clover are grown together on the same plot, experience has shown that they can be modeled as if they develop fairly independently up to the time the mature plants set their seed. But the seed numbers set depend on the densities of both species at this time. Thus they may be reduced below the monoculture yields, or the strains could cooperate and each produce higher yields, or perhaps one strain benefits at the expense of the other. Both strains will also be subject to competition from invading (volunteer) species. Seed yields in binary mixtures are taken to be described by the de Wit replacement model which is based on the law of mass action from chemistry. Until recently there does not seem to have been much theoretical support for this description, but it does seem well supported by field observations.

The life history of clover can be characterised by many parameters, such as age dependent seed softening rates and survival probabilities of seedlings and seed, etc. In Western Australia, agronomists have done much experimental work relating to these properties, including a seven-year-long study in which six clover strains were grown together as three binary mixtures. Several years ago R. A. Maller was approached by an eminent local agronomist and asked if it would be possible to answer the four basic questions outlined above by means of appropriate mathematical modeling of the dynamics of clover growth. A further approach by him to A. G. Pakes resulted in this monograph; in essence it is a detailed mathematical study of the above agronomical questions. It formulates and analyses four specific models which are differentiated on the basis of experimental exigencies, namely, on the detail in which it is possible to differentiate and count the differing seed types in the seed bank.

These models have several nice features. First, it turns out that three of the four differing experimental situations give rise to the same mathematical structure, while the remaining situation produces a further simplification giving a different, but more tractable, mathematical structure. Second, although the agronomic situation is described in terms of a very large number of parameters, about fifty for one case, these aggregate in the mathematical models, thus giving descriptions in terms of considerably fewer parameters. It transpires that the most significant features of the models can be characterized in terms of a small number of derived constants. Third, the model equations are non-linear iterations, a currently fashionable mathematical topic. They have not previously arisen in the literature and are sufficiently complicated that it is not possible to explicitly calculate their time dependent behaviour, yet they are simple enough to permit the construction of a detailed picture of their qualitative behavior. Of course this means that the models don't show chaotic behavior, almost *de rigueur* these days. (This would very likely change if other density-dependent effects, such as self thinning, were incorporated.) However, they do contain all the factors deemed important by the agronomists we have consulted.

The book begins with an introductory chapter which sets the scene with a non-mathematical outline of the growth dynamics of clover. Formulation of the models from these dynamics is quite a long task and Chapter 2 is devoted to this. There are two main models, Models 1 and 2, having the same mathematical structure. This structure is in the form of a pair of non-linear discrete convolution equations. A special case of Model 1 yields a simpler model, called Model 1A, which has the form of a pair of first order non-linear difference equations.

The analysis of Model 1A is presented in Chapters 3–5. Chapter 3 introduces this analysis by imbedding Model 1A in a more general family of models, called Model G (for general), thus giving a situation much closer in spirit to contemporary work on population mathematics. Our exposition is couched in terms of the concepts and jargon of elementary dynamical systems theory. We summarize what is needed in Chapter 3. The chapter ends with a fairly long review of recent theoretical work about modeling the growth of annual plants having seed banks. This addresses three themes. The first concerns literature about first order pairs of difference equations of general form, and hence we exclude the massive literature dealing with systems having a specific parametric form and exhibiting chaotic behaviour – this is not relevant to the work described here. The second theme is to do with the evolution of dormancy. The third theme is to do with the specific models that have appeared elsewhere in the literature dealing with the dynamics of plant populations having a seed bank. This section should provide a context for the main study, and it lets the reader see which issues of potential importance to plant growth have been ignored here, and how they have been handled by other workers.

Model G can have any number of equilibria, but we always assume there is at most one interior equilibrium, because this is what happens for Model 1A. Chapter 4 presents a detailed analysis of the case where there is no interior equilibrium. There are two boundary equilibria, corresponding to extinction of one of the strains, and we show that one of these is always globally stable and attracting and give precise conditions for determining which it is. The proof uses invariance arguments, that is, certain regions defined by the isoclines cannot be exited once they are entered. The final configuration of the orbits can be determined from an examination of the rate of their approach to the equilibria. A fairly thorough picture of the shape of the orbits of Model 1A is built up in a section which is not required for later sections. Since this is algebraically complicated, the reader may wish to skip this section on the first reading. The last section of Chapter 4 treats a degenerate form of Model G which cannot correspond to any possible configuration of Model 1A.

Chapter 5 covers similar topics when there is a single interior equilibrium. In the case of Model 1A, a Liapunov function is used to give an alternative proof that the interior equilibrium is stable and attracting.

Models 1 and 2 are treated in Chapter 6. These models give rise to a pair of difference equations of order equal to the current time variable. If ungerminated seeds have a finite life time (termed "finite memory") then the resulting system can be imbedded in a dynamical system moving in a higher dimensional Euclidean space. The long term out-come of this system is determined by borrowing methods used in the theory of branching processes, and this resolves the issue for clover outcomes. Again, attention is given to rates of convergence. Finally a local analysis is given for the biologically irrelevant case of infinite memory. Throughout most of this chapter we assume that the age dependent seed softening rates have a finite sum, when summed over all possible age classes, but we investigate a couple of special cases where this condition is relaxed. This has no biological significance, but it is mathematically interesting to find that the qualitative behavior of the model is unchanged, though there is a quantitative change in that convergence to equilibria occurs much more slowly.

Chapter 7 opens with a recapitulation of the most important theoreti-cal findings from the previous chapters and then discusses their signifi-cance for data obtained from field experiments for several binary mixtures of clover. All the field data indicate that one strain will drive the other to extinction, but the convergence rate results show that this will not be discernible in some of the mixtures. For example, in one it will take some 600 years for the losing strain to be effectively eliminated. The models allow one to determine which parameters are most important in affecting the speed of establishment of the winning strain. This is important to agronomists, and our results show that the folk belief that softening rates are always most important is unfounded. Some of this chapter is devoted to explaining the origin of the data, which comes from diverse sources.

Chapter 8 points to a possible direction of future research whilst restricting discussion to a single strain of subclover. Here a much more detailed model is introduced to follow the processes of cropping, whereby pasture is removed to allow production of a crop, such as wheat, and tillage, whereby clover seed is redistributed underground. These pro-cesses potentially apply to (binary) mixtures of clovers also, but it turns out that a reasonably realistic model even for a single strain is a research topic in itself, and Chapter 8 is only an introduction to this subject. Our future work will be directed towards synthesising the models of Chapter 8 with the mixture models.

This book is written in a mathematical style, that is, theorem–proof–discussion. We believe that this caters better for readers with differing interests, and in particular it clearly separates the proofs, wherein lie most of the technicalities, from the remainder. Thus the results and their consequences can be gleaned without having to plough through the proofs and the latter are always easily identified. The book can be read either by mathematicians or mathematically literate biologists.

Most of the mathematics concerns convergence properties of sequences and series, but some parts require knowledge of discrete renewal equations and iteration of monotonic functions. All of it should be accessible to anyone who has coped with a course in advanced calculus, although most of the formulation and discussion in the book is accessible with less of this background. The major points will probably be most appreciated by readers having some knowledge of the biology of pasture crops. In particular, in some parts we make frequent references to the paper Rossiter, Maller, and Pakes (1985). This contains much detail about field aspects and discusses the implications of field data using Model 1A. We suggest that readers obtain a copy of this paper for consultation when necessary.

Any self-respecting preface should end with an appreciation to all who have contributed to the writing of the work. We are indeed indebted to Reg Rossiter, whose passion for subclover led to the research described herein, and we thank him for his advice and interest in our project since its inception in 1983. R.A.M. thanks also Graham Taylor for his help and advice, particularly in connection with the data used in Chapter 8 for the single strain model. Finally, we thank Joyce, Kath, Allison, Cathy, Lyn, and Ingrid, the typists in the UWA Department of Mathematics, who worked on portions of the manuscript in its various incarnations.

Perth, December 1988 Anthony G. Pakes
Ross A. Maller

1

Introduction

1.1   Introduction

Subterranean clover (*T. subterraneum L*) based pastures have proved outstandingly successful in Australia. The area of pasture in which subterranean clover is estimated to form the principal legume component exceeds 12×10^6 ha in Australia, with over half of this in Western Australia alone. Smaller, but substantial, areas occur in Western USA and in the Mediterranean, which is its area of natural distribution. The success of subterranean clover in establishment and persistence is attributed at least in part to the large number of varieties (strains) within the species, which enhance its adaptability to a wide range of ecological environments. In Australia, subterranean clover is often sown into old pastures, which are dominated by one or more other strains of subterranean clover, and by various other annual or perennial species. Other pastures seem to be a balanced mixture of two or more clovers.

Our interest in this monograph is in developing models to describe the growth of, and predict the eventual success of, strains of subterranean clover grown in binary mixtures. Our major motivation for such a study appears in Rossiter (1974), and stems from the need to know when an undesirable strain (for example, one containing toxic chemical compounds) may be replaced by a more desirable one; conversely, a new strain should be capable of maintaining itself against serious contamination by invading strains. Added to this is the possibility of describing the persistence of two (or more) desirable strains in dynamic equilibrium in a pasture. Many such binary mixtures are well established in Western Australia, see Rossiter (1966).

The mathematical models given in this monograph are an attempt to formalise knowledge gained over many years of observation and experimentation with competing strains of subterranean clover in the south-west of Western Australia. But, although based on specific and

1

localised data, and benefiting greatly thereby in insight gained from the necessity of being applicable to the real agronomic problems of the area, the models are quite general mathematically and can potentially be applied to many plant strains, and also species sharing the major characteristics of subterranean clover. Hence the general title of this monograph. The methods may also have applications to competing systems of species, other than plant species; see Section 4 of Chapter 3.

Subterranean clover is a winter-growing annual plant, that is, individual plants die at the end of each year's growing season and the strain regenerates from seed produced mainly in the previous year. An important feature of the annual seed set by subterranean clover, shared by some related plant species (for example, annual medics or *Medicago* spp.) is that the seed is *hard*, that is, impermeable to water. Over the summer a proportion of this seed *softens*, and it is this seed only which germinates in the next growing season. The seed which does not soften remains as *residual hard seed*, and is carried into the following year as part of the total seed pool. A further proportion of this seed softens over the next summer and germinates in the following growing season. In some species, for example, grasses, this seed carryover occurs as dormant seed.

The hardseededness of subterranean clover has been the subject of intensive study by Taylor, Rossiter, and Palmer (1984) and others, and many important and curious features have been discovered. Chief among these, as far as our present models are concerned, is that hard seed may soften in an age-dependent way, that is, hard seed left from year N may soften at a faster rate than seed set from years later than N. Some strains may not show this effect, that is, seed may soften at approximately the same rate regardless of its age, but nevertheless the possibility of different rates should be allowed for in a serious attempt to model hardseededness of subterranean clover.

We also need to take into account the fact that, subterranean clover seed occurs initially in *burrs*, but that after some time (due to aging or trampling by stock) the seed may free itself from the burr and become *free seed*. This process, which may occur at different rates depending on the age of the burr, is important because free seed softens at a faster rate than burr seed. There are two possible ways of treating burr and free seed, which make up two major classes of models: in models of type 1, residual burr and free seed are only differentiated each year at the time of seed softening, whereas in models of type 2, burr and free seed of all ages are accounted for separately. Model 2 is more realistic, but Model 1 is important since often in practice only the bulked burr and free seed is observed. The essential mathematics of both models turns out to be the same.

From the basic considerations of hardseededness and burr/free seed we can start to build up a deterministic life cycle or life history model of a single subterranean clover strain. Such kinds of models of course have long been used in animal and plant biology to describe the population dynamics of species, that is, the way individuals of generation n give rise to individuals of generation $n + 1$, and the rates at which the transition takes place. For a number of examples, see Sagar and Mortimer (1976). Life history models have been analysed mathematically from many points of view, and Leslie's approach (Pollard (1973), for example), is fundamental to many of these.

For mixtures of competing strains we find that Leslie matrices are of little use when we add the essential ingredient needed to describe the *interaction* of the two strains. The mechanism for this was first described for subterranean clover by Rossiter and Palmer (1981) who showed that the seed yield of each of two strains grown in a binary mixture could be well predicted from the proportion of adult plants of each strain in the mixture by means of a de Wit replacement curve. This is a simple, intuitively well founded and biologically well based (for example, Trenbath (1976)) formulation which is nevertheless *non-linear*, and it is this aspect of our models which provides many of their characteristics. The actual mechanism of plant competition depends on many factors which are not relevant to us, as long as the result is well described by the de Wit series.

A class of models derived from the above considerations was formulated and used to describe experimental data reported in Rossiter et al. (1985). That paper only reports and investigates the applicability of the simplest of those models (denoted later as Model 1A) to the data. The purpose of the present monograph is to provide the mathematical formulation and asymptotic analysis of the whole class of models. These aspects of the models provide a knowledge of the long term behavior of the mixture and allow an investigation of the way the various parameters of the model affect the eventual success or otherwise of a strain. By contrast, Rossiter et al. (1985) give a detailed discussion of Model 1A, in which the dynamic behavior of three actual binary mixtures is compared with the mathematical predictions from the model.

The organisation of the present monograph is as follows. For the remainder of this chapter we give a general introduction to the formulation of the models. In Chapter 2 a detailed mathematical formulation is given. Chapters 3, 4 and 5 give an asymptotic analysis of Model 1A;

Chapter 6 analyses the more general Models 1 and 2; and in Chapter 7 the models are discussed in relation to field data obtained from experimental plots of binary mixtures of different strains of subterranean clover. In Chapter 3 we have also included a review of earlier theoretical work on difference equation models of populations of annual plants having seed banks. Finally, in Chapter 8 we discuss a topic which is somewhat apart from the main theme of the monograph. Here we present some difference equation models of the dynamics of single strain clover pastures subject to cropping and tillage. This chapter is largely independent of the remainder, although it would be useful to read this and the next chapters for orientation.

We close the present section with a discussion of some of the shortcomings of the model. Our major assumption will be that the basic parameters of the models, such as softening rates, remain constant from year to year. In practice, of course, these vary with seasonal conditions, so what is required is a *stochastic* version of the models. We hope to discuss some aspects of these elsewhere. The models could also be generalized to include some aspects of *tillage*, by which means the clover seed is distributed below the ground when pasture is grown in rotation with cereals. A single strain model with tillage has been formulated and analysed in Chapter 8 and could in principle be extended to mixture models. Our analysis there also incorporates *density dependence* in the seed production, which we assume is absent in the mixture models for reasons given in Rossiter *et al.* (1985). This could also be included in our mixture models with a little extra difficulty.

1.2   General description of the models

In Figure 1.1, a simple life history model for a single strain of subterranean clover is schematized. At the end of winter in year n, new seed is set by the mature plants. Together with residual hard seed this forms a total seed pool, which is subject to losses (by grazing animals, premature germination, and so forth) over summer, so that only a proportion P_S of it remains at the beginning of winter in year $n + 1$. By this time also a proportion $P(S)$ of the seed has softened, while $P(H) = 1 - P(S)$ of it remains as residual hard seed. The soft seed germinates and gives rise to established plants with probability P_E and adult plants with probability P_A. These produce new seed (at a rate which we will not specify at this stage) and so complete the cycle. We also take into account the fact that volunteer (non-clover) species may enter the pasture at a rate P_V which may depend on time.

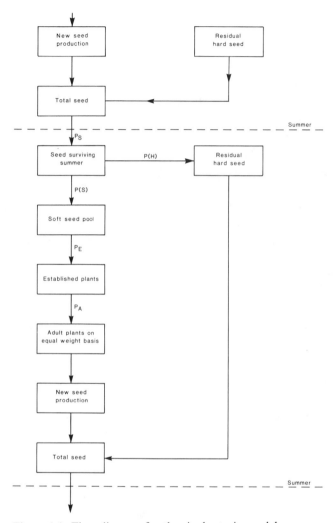

Figure 1.1. Flow diagram for the single strain model.

To formulate the mixture model, we think of two strains following this cycle in parallel. We assume they interact only at the stage of new seed production, and that they do this by means of a de Wit replacement series. The de Wit (1960) series is well established in the literature of competing plant species and has been described many times. A description in the notation of the present paper occurs in Rossiter et al. (1985) which we will not repeat here, but just mention that the predicted seed yields of Strains 1 and 2 are given, respectively, by

$$Y_1 = \frac{M_1 P_V k_{12} z}{(k_{12} - 1)z + 1}$$

and

$$Y_2 = \frac{M_2 P_V k_{21}(1 - z)}{(k_{21} - 1)(1 - z) + 1},$$

where z is the proportionate adult plant yield of Strain 1 on a weight basis, and $1 - z$ is that of Strain 2. The constants M_1 and M_2 are the monoculture yields of the two strains, that is, their yields when grown as single strains in isolation. Finally, k_{12} and k_{21} are the "de Wit constants" or "relative crowding coefficients," and P_V is a discount factor for volunteer species. We refer the reader to Braakhekke (1980), especially Chapter 4, for a detailed discussion of the de Wit model.

A flow diagram for the simplest of the mixture models (called Model 1A in the next section) is given in Rossiter et al. (1985) and consists simply of Figure 1.1 together with an imagined mirror image scheme for the second strain, with an interaction indicated at the adult plant to new seed stage.

As outlined previously, our more general models include the age-dependent softening of free and burr seed and, for one class of models, the age dependent rate at which burr seed becomes free. Thus the models have a "history," in that the seed pools of various ages are distinguished in order to determine their contribution, via soft seed, to the current year's established plants.

The two main classes of models depend on the way burr and free seed are treated. In Model 1, burr and free seed are only differentiated for the purposes of seed softening; there are reasons other than mathematical convenience for this, since often all that is observed experimentally is bulked burr and free seed of all ages for each strain. In principle, however, it is possible to distinguish free and burr seeds of various ages, and so our second class of models, Model 2, keeps track of burr and free seed of all ages.

Within these two classes of models we distinguish various sub-models, which are simplifications of the main models when one or another process is assumed not to operate. For example, Model 1A, which is formulated using the concept of *residual hard seed*, turns out to be a submodel of Model 1 in which seed of all ages contributes equally to the current year's soft seed pool. Model 2A is a submodel of Model 2 in which burr seed does not become free after the first year.

The mixture models can now be described in general terms as follows: over summer, a proportion of seed (P_S and Q_S for the two strains) survives and the seed becomes one year older. (Notation will be explained in the next chapter.) It is composed of burr and free seed which soften at different rates to form a soft seed pool. Also contributing to this pool is seed which softens from two-, three-, ... year old seed. Seed which remains hard forms pools of older seed in later years. The soft seed under consideration now germinates and gives rise to established plants with probabilities P_E, Q_E and these survive to adulthood with probabilities P_A, Q_A. Because of strain differences in seed size, a further factor W is introduced at this stage for Strain 1 only, to place the adult plant yield on an equal weight basis. Now the de Wit model is used to predict the seed yield for each strain, which, when discounted for volunteer species intrusion, yields the current seed production at the beginning of summer. This completes the year's cycle.

To follow this process in detail for Model 1, we refer the reader to the flow diagram of Figure 1.2. This shows only one half of the model, that for Strain 1. Stain 2 can be included by envisaging the mirror image of Figure 1.2 placed on the left hand side of the figure, with Q replacing P throughout. The two strains interact only in the de Wit part of the model. In Figure 1.2 the arrows denote transitions between the states written in the boxes. Alongside each arrow is the probability of this transition, i.e. the rate at which it takes place.

It should be noted in Figure 1.2 that burr and free hard seed are lumped together each year so that we only carry pools of two-, three-year-old seed, and so forth, into later years. Then each year these pools are separated into burr and free seed again in the proportions $P(B|B)$ and $P(F|B) = 1 - P(B|B)$.

Thus in Model 1 (and in its submodel 1A), the proportion of free seed is always exactly $P(F|B)$. Model 1 was derived on the assumption that we only observe a bulked sample of seed which consists of free and burr seed in a certain proportion.

In reality, seed, once free, remains free while burr seed becomes free at a rate which may depend on its age. Model 2 incorporates these features, which distinguish it from Model 1: Here we keep burr and free seed of all ages separate, and allow a transition from burr to free seed at any age. The flow diagram for this model is given in Figure 1.3. The greater reality of Model 2 is gained at the expense, however, of not knowing explicitly the proportion of free seed in the system. In fact, the proportion is time dependent and must be calculated each year. For this model, $P_1(F|B)$ refers only to the rate at which *one-year-old seed becomes free*, whereas in Model 1 this rate is also the proportion of free seed in the whole system.

We conclude this section with a general description of how we calculate the soft seed pool at the break of season; the details are given later in the next chapter. Under both types of models the seed produced in year n, say, becomes one-year-old hard seed after a proportion has

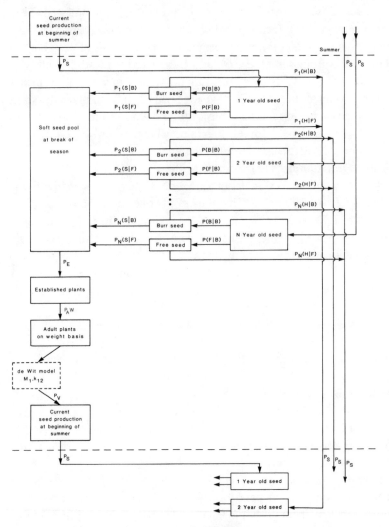

Figure 1.2. Flow diagram for Model 1.

been removed by summer losses. Of the one-year-old seed, a further proportion is lost over summer and by softening in year $n + 1$, leaving a residuum of two-year-old seed. The process continues to give three-, four-, ... year old seed.

Initially all seed produced is burr, but some of this becomes free each year. Burr and free seed soften at different rates and, under Model 2, burr seed becomes free as it ages, so separate calculations are made for burr and free seed.

The proportions of the current year's production (year n, say) of one strain which survive one, two, three, ... years can be visualized as

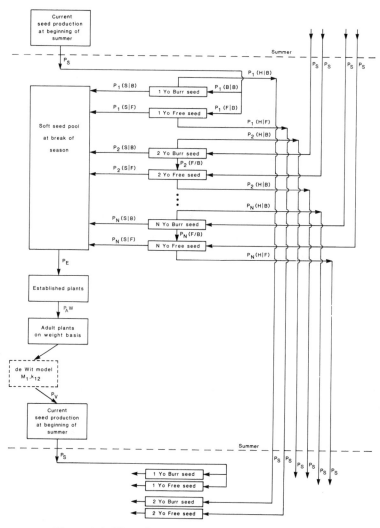

Figure 1.3. Flow diagram for Model 2.

occurring in the following scheme:

year n current production + 1 year old

$n+1$ current production + 1 year old + 2 years old

$n+2$ current production + 1 year old + 2 years old + 3 years old

$n+3$ current production + 1 year old + 2 years old + 3 years old

current production + 1 year old + 2 years old + 3 years old

etc.

Thus, in year n, seed of age one, two, ... $n - 1$ years is present and potentially able to contribute to the soft seed pool for that year. The rates at which the transitions in the above diagram take place are calculated in the next chapter.

2

Mathematical formulation of the models

2.1 Introduction

Our models can be regarded as extensions of the de Wit (1960) model for seed production of competing strains when we include the further information on seed survival, establishment, and so forth, necessary to give a description of the annual cycle of growth of the mixture. de Wit himself gave a simple deterministic population model in which he assumed that the seed produced according to his replacement model was resown the following year, and the resultant plants (assuming no losses) produced seed again according to the replacement model.

In this chapter we give the mathematical formulation of the models in terms of the basic observable parameters, and discuss the relation of the models to each other and to the various submodels. We also prove a result of some practical importance which shows that the longer term behaviour of some models of Type 2 is not affected if the age dependent softening rates are replaced by an average softening rate.

Notation presents a special problem due to the large number of parameters, and the need to distinguish them for the two strains. We have tried to strike a balance between readability and generality. Basically the system is as follows. The various survival and establishment rates, which in fact are probabilities of survival and establishment, will be denoted for the first strain by a subscripted P. For example P_S represents the probability of summer survival for Strain 1.

We also need to introduce conditional probabilities to describe, for example, the rate at which burr and free seed soften. These would be written as $P(S|B)$ and $P(S|F)$ respectively. But in some models these rates are dependent on age of the seed, so we need to introduce a subscript: $P_i(S|B)$ is the rate at which i-year-old burr seed of the first strain softens, $i = 1, 2, \ldots$ Further probabilities are defined by complementation: $P_i(H|B)$ is the probability of i-year-old burr seed of Strain 1 remaining hard; $P_i(H|B) = 1 - P_i(S|B)$.

For the second strain, rather than introduce another subscript, we use the letter Q to denote the same probabilities. Thus Q_S is the probability that a seed of Strain 2 survives summer. An exception to this scheme is made for the de Wit component of the model, where we follow fairly closely de Wit's original notation. Also, since volunteer species are assumed to affect both strains equally, we only use a single symbol, P_V, for these.

We need a formidable array of symbols to describe the most general models. For some of the submodels, however, including Model 1A, the number reduces significantly. In Table A.1 we list the symbols for Models 1 and 2, the most general cases of the two types, and for Models 1A and 2A, which are the most important submodels of each type. Only the symbols for Strain 1 are given, except for the de Wit constants, which are given for both. The other symbols for Strain 2 are defined by replacing P by Q everywhere in Table A.1. Estimates of these probabilities for the experimental mixtures of Rossiter et al. (1985) are given in Chapter 7.

2.2 Formulation of Model 1

Consider the seed produced by Strain 1 in a given year which has survived over summer and hence is reduced to a proportion P_S of the original. This we call one-year-old seed. Of this, a proportion $P(F|B)$ is free and $P(B|B) = 1 - P(F|B)$ is burr. Of the burr seed a proportion $P_1(S|B)$ softens and of the free seed, $P_1(S|F)$ softens. What remains hard of this seed goes to make up next year's two-year-old seed. This has survived two summers so it will be the proportion

$$P_S^2\{[1 - P_1(S|B)]P(B|B) + [1 - P_1(S|F)]P(F|B)\}$$

$$= P_S^2[P_1(H|B)P(B|B) + P_1(H|F)P(F|B)]$$

$$= P_S^2 P_1(H) \tag{2.1}$$

of the original. Here we defined $P_i(H|B) = 1 - P_i(S|B)$, $P_i(H|F) = 1 - P_i(S|F)$, and $P_i(H) = P_i(H|B)P(B|B) + P_i(H|F)P(F|B)$ is the probability of i-year-old seed remaining hard. *Note that the burr and free components of this residual seed are not identified.*

Of this two-year-old seed, a proportion $P_S^2 P_1(H)P(B|B)$ is burr and $P_S^2 P_1(H)P(F|B)$ is free, and these soften at the rates $P_2(S|B)$ and $P_2(S|F)$ respectively, leaving hard the proportion

$$P_S^2\{[1 - P_2(S|B)]P_1(H)P(B|B) + [1 - P_2(S|F)]P_1(H)P(F|B)\}$$

$$= P_S^2 P_2(H)P_1(H). \tag{2.2}$$

Repeating this process, we see that the i-year-old hard seed is

$$H_i = P_S^i \prod_{k=1}^{i-1} P_k(H), i \geqslant 1. \tag{2.3}$$

(Here and throughout this monograph we make the convention that

$$\prod_{k=j}^{i} = 1 \text{ if } i < j.)$$

The soft seed produced from this is, for Strain 1, in the proportion

$$S_i = P_S^i P_i(S) \prod_{k=1}^{i-1} P_k(H), i \geqslant 1, \tag{2.4}$$

where $P_i(S) = 1 - P_i(H)$.

Let x_n be *the new seed produced by Strain 1 at the beginning of summer in year n*, and y_n that for Strain 2. Then the soft seed pool at the break of season in year n, $n \geqslant 2$, for Strain 1 is

$$\sum_{i=1}^{n-1} S_i x_{n-i} = \sum_{i=1}^{n-1} P_S^i P_i(S) \prod_{k=1}^{i-1} P_k(H) x_{n-i},$$

where x_1 is the amount of seed with which Strain 1 starts. For Strain 2, formulae (2.1) and (2.2) hold with P replaced by Q, so the soft seed pool for this strain is

$$\sum_{i=1}^{n-1} T_i y_{n-i} = \sum_{i=1}^{n-1} Q_S^i Q_i(S) \prod_{k=1}^{i-1} Q_k(H) y_{n-i}. \tag{2.5}$$

(Here and throughout the monograph we follow the convention of denoting quantities for Strain 2 by the letter of the alphabet following that for Strain 1; hence S_i becomes T_i when Q replaces P in (2.4).)

These soft seeds germinate and give rise to established plants and adult plants with probabilities $P_E P_A$ and $Q_E Q_A$; if we put Strain 1 on the same weight basis as Strain 2 by multiplying by W, we see that the proportionate yield of Strain 1 relative to Strain 2 at this stage is

$$z = P_E P_A W \sum_{i=1}^{n-1} S_i x_{n-i} / D \tag{2.6}$$

while that for Strain 2 is

$$1 - z = Q_E Q_A \sum_{i=1}^{n-1} T_i y_{n-i} / D \tag{2.7}$$

where

$$D = P_E P_A W \sum_{i=1}^{n-1} S_i x_{n-i} + Q_E Q_A \sum_{i=1}^{n-1} T_i y_{n-i}.$$

At this stage we are ready to apply the de Wit model; for a formulation of what we need, see Section 2 of Chapter 1. The seed produced at the beginning of summer in year n, which is just x_n, is given by

$$x_n = \frac{M_1 k_{12} P_V z}{(k_{12} - 1)z + 1}. \tag{2.8}$$

Since z depends on $x_{n-1}, x_{n-2}, \ldots$ and on $y_{n-1}, y_{n-2}, \ldots$, (2.8) is a mixed difference equation of (potential) order n for x_n. Similarly

$$y_n = \frac{M_2 k_{21} P_V (1 - z)}{1 - (1 - k_{21})(1 - z)}. \tag{2.9}$$

An interesting feature of the models, which will be shown in Chapter 6 to be related to the long term outcome of the populations, is the total amount of soft seed ever produced. We define quantities A, B which are proportional to these by

$$A = P_E P_A P_V M_1 W \sum_{i=1}^{\infty} S_i$$

$$= P_E P_A P_V M_1 W \sum_{i=1}^{\infty} P_s^i P_i(S) \sum_{k=1}^{i-1} P_k(H) \tag{2.10}$$

and

$$B = Q_E Q_A P_V M_2 \sum_{i=1}^{\infty} T_i.$$

2.3   Formulation of Model 1A

Referring to the flow diagram for this model (Figure 3 of Rossiter et al. (1985)) we have, if x_{n-1} and y_{n-1} are the seed productions in year $n - 1$ by the two strains, and r_{n-1} and s_{n-1} are the residual hard seed quantities carried over in year $n - 1$ (that is, seed that does not soften in year $n - 1$) then, in year n, $P_S P(S)(x_{n-1} + r_{n-1})$ seed of Strain 1 survives summer. Of this, $P_S P(S)(x_{n-1} + r_{n-1})$ softens and

$$r_n = P_S P(H)(x_{n-1} + r_{n-1}) \tag{2.11}$$

remains hard. Here $P(S) = P(S|B)P(B|B) + P(S|F)P(F|B) \equiv 1 - P(H)$. Similarly for Strain 2. Applying the de Wit model for the adult plants on a weight basis and discounting for volunteers gives

$$x_n = \frac{M_1 P_V k_{12} z}{(k_{12} - 1)z + 1}, \quad y_n = \frac{M_2 P_V k_{21}(1 - z)}{1 - (1 - k_{21})(1 - z)} \tag{2.12}$$

where

$$z = P_E P_A W P(S) P_S (x_{n-1} + r_{n-1})/D,$$

$$D = P_E P_A W P(S) P_S (x_{n-1} + r_{n-1}) + Q_E Q_A Q(S) Q_S (y_{n-1} + s_{n-1}).$$

So the maximum seed pool for Strain 1 is

$$t_n = x_n + r_n$$

$$= \frac{k_{12} M_1 P_V P_E P_A WP(S) P_S t_{n-1}}{k_{12} P_E P_A WP(S) P_S t_{n-1} + Q_E Q_A Q(S) Q_S u_{n-1}} + P_S P(H) t_{n-1},$$

(2.13a)

and for Strain 2 the maximum seed pool is

$$u_n = y_n + s_n$$

$$= \frac{k_{21} M_2 P_V Q_E Q_A Q(S) Q_S u_{n-1}}{P_E P_A WP(S) P_S t_{n-1} + k_{21} Q_E Q_A Q(S) Q_S u_{n-1}} + Q_S Q(H) u_{n-1}.$$

(2.13b)

The difference equations for the new seed productions, x_n and y_n, are of order $n - 1$ and can be obtained as follows. Using (2.11),

$$r_n = P_S P(H)(x_{n-1} + r_{n-1})$$

$$= P_S P(H) x_{n-1} + P_S^2 P^2(H)(x_{n-2} + r_{n-2})$$

$$\vdots$$

$$= \sum_{i=1}^{n-1} P_S^i P^i(H) x_{n-i},$$

and substituting in (2.12) gives, for the new seed productions,

$$x_n = \frac{k_{12} M_1 P_V P_E P_A WP(S) P_S \left(x_{n-1} + \sum_{i=2}^{n-1} P_S^{i-1} P^{i-1}(H) x_{n-i} \right)}{k_{12} P_E P_A WP(S) P_S \left(x_{n-1} + \sum_{i=2}^{n-1} P_S^{i-1} P^{i-1}(H) x_{n-i} \right) + Q_E Q_A Q(S) Q_S \left(y_{n-1} + \sum_{i=2}^{n-1} Q_S^{i-1} Q^{i-1}(H) y_{n-i} \right)}$$

and

$$y_n = \frac{k_{21} M_2 P_V Q_E Q_A Q(S) Q_S \left(y_{n-1} + \sum_{i=2}^{n-1} Q_S^{i-1} Q^{i-1}(H) y_{n-i} \right)}{P_E P_A WP(S) P_S \left(x_{n-1} + \sum_{i=2}^{n-1} P_S^{i-1} P^{i-1}(H) x_{n-i} \right) + k_{21} Q_E Q_A Q(S) Q_S \left(y_{n-1} + \sum_{i=2}^{n-1} Q_S^{i-1} Q^{i-1}(H) y_{n-i} \right)}$$

(2.14)

Referring to (2.8) and (2.9), it is not difficult to check that exactly these equations are obtained by putting $P_i(S) = P(S)$, $P_i(H) = P(H)$, $Q_i(S) = Q(S)$, $Q_i(H) = Q(H)$. Thus: *Model 1A is a submodel of Model 1 in which seed of all ages is assumed to soften at the same rate.*

The quantities A and B given by (2.10) become, for Model 1A,

$$A = M_1 P_V P_E P_A W \sum_{i \geqslant 1} P_S^i P(S) P^{i-1}(H) = \frac{M_1 P_V P_E P_A W P(S) P_S}{1 - P_S P(H)}$$

and

$$B = \frac{M_2 P_V Q_E Q_A Q(S) Q_S}{1 - Q_S Q(H)}. \tag{2.15}$$

2.4 Formulation of Model 2

Consider the seed produced by Strain 1 in a given year which has survived over summer, that is, one-year-old seed. This is reduced to a proportion P_S of the original. Of this a proportion $P_1(B|B)$ is burr seed and $1 - P_1(B|B) = P_1(F|B)$ is free seed. Thus one-year-old burr seed is a proportion $P_S P_1(B|B)$ of the original and one-year-old free seed is a proportion $P_S P_1(F|B)$ of the original.

Of the burr seed, a proportion $P_1(S|B)$ softens in the current year and a proportion $1 - P_1(S|B) = P_1(H|B)$ remains hard. Thus the remaining one-year-old hard burr seed is $P_S P_1(H|B) P_1(B|B)$, while that which softened from one-year-old seed was $P_S P_1(S|B) P_1(B|B)$.

Over the next summer the hard burr seed decreases by a further proportion P_S and becomes two-year-old seed. A further proportion of this, $P_2(F|B)$, becomes free, while $P_2(S|B)$ of it softens. Thus the remaining two-year-old hard burr seed is

$$P_S^2 P_1(H|B) P_2(H|B) P_1(B|B) P_2(B|B)$$

while seed which softened from two-year-old burr seed was

$$P_S^2 P_2(S|B) P_1(H|B) P_1(B|B) P_2(B|B).$$

Repeating this process we see that the i-year-old hard burr seed remaining after softening in year i is in the proportion

$$H_i(B) = P_S^i \prod_{k=1}^{i} P_k(H|B) \prod_{k=1}^{i} P_k(B|B) \tag{2.16}$$

while seed which softened from i-year-old burr seed was in the proportion

$$S_i(B) = P_S^i P_i(S|B) \prod_{k=1}^{i-1} P_k(H|B) \prod_{k=1}^{i} P_k(B|B). \tag{2.17}$$

To calculate the free seed remaining after softening, we proceed as follows. The free seed remaining after softening in year $i + 1$ will have a component of free seed which has remained hard from the ith year, survived the summer of the ith year, and not softened in year $i + 1$.

This will be the proportion

$$P_S P_{i+1}(H\,|\,F) H_i(F)$$

where we define

$H_i(F)$ = free seed remaining after softening in year i.

To this must be added the burr seed which becomes free in the $(i+1)^{\text{th}}$ year. This is burr seed which remains hard from the ith year, survives the summer of the $(i+1)^{\text{th}}$ year, then becomes free. Thus, from (2.16), it is in the proportion

$$P_S P_{i+1}(F\,|\,B) H_i(B) = P_S^{i+1} P_{i+1}(F\,|\,B) \prod_{k=1}^{i} P_k(H\,|\,B) \prod_{k=1}^{i} P_k(B\,|\,B).$$

(2.18)

Hence the proportion of free seed remaining in year $i+1$, which is $H_{i+1}(F)$, is equal to

$$H_{i+1}(F) = P_S P_{i+1}(H\,|\,F) H_i(F)$$

$$+ P_S^{i+1} P_{i+1}(F\,|\,B) \prod_{k=1}^{i} P_k(H\,|\,B) \prod_{k=1}^{i} P_k(B\,|\,B).$$

This is a recursive relation for $H_i(F)$ of the form $H_{i+1}(F) = a_i H_i(F) + b_i$, with $H_1(F) = P_S P_1(H\,|\,F) P_1(F\,|\,B)$. Solving this, we obtain

$$H_i(F) = P_S^i \prod_{k=1}^{i} P_k(H\,|\,F) P_1(F\,|\,B)$$

$$+ P_S^i \sum_{k=1}^{i-1} P_{k+1}(F\,|\,B) \prod_{j=1}^{k} P_j(H\,|\,B) P_j(B\,|\,B) \prod_{j=k+1}^{i} P_j(H\,|\,F)$$

(2.19)

where, in addition to our convention $\prod_{j=k}^{l} = 1$ when $k > l$, we understand $\sum_{k=1}^{0} = 0$.

In a similar way we find the seed which softened from i-year-old free seed to be

$$S_i(F) = P_S^i P_i(S\,|\,F) \prod_{k=1}^{i-1} P_k(H\,|\,F) P_1(F\,|\,B)$$

$$+ P_S^i P_i(S\,|\,F) \sum_{k=1}^{i-1} P_{k+1}(F\,|\,B) \prod_{j=1}^{k} P_j(H\,|\,B) P_j(B\,|\,B) \prod_{j=k+1}^{i-1} P_j(H\,|\,F).$$

(2.20)

The soft seed contribution in any year from i-year-old seed is, as a proportion of the seed produced i years ago,

$$S_i = S_i(B) + S_i(F). \tag{2.21}$$

Similarly, for hard seed we have

$$H_i = H_i(B) + H_i(F). \tag{2.22}$$

So far we have been working with probabilities or proportions of seed, and always for Strain 1. To put this on a yield basis, let x_n *be the new seed produced by Strain 1 at the beginning of summer in year n.* Let y_n be the same quantity for Strain 2. We now calculate x_n and y_n.

The *soft seed pool at the break of season* in year n is made up of the soft seed contributions from seed of ages $1, 2, \ldots, n-1$ years, which occur in proportions $S_1, S_2, \ldots, S_{n-1}$ given by (2.21). Thus the soft seed pool for Strain 1 is

$$\sum_{i=1}^{n-1} S_i x_{n-i} \tag{2.23}$$

and that for Strain 2 is

$$\sum_{i=1}^{n-1} T_i y_{n-i}, \tag{2.24}$$

where T_i is defined in exactly the same way as S_i but with Q replacing P throughout.

The adult plants produced in year n are, on an equal weight basis,

$$P_E P_A W \sum_{i=1}^{n-1} S_i x_{n-i} \quad \text{and} \quad Q_E Q_A \sum_{i=1}^{n-1} T_i y_{n-i},$$

and from these we can calculate the *new seed production* for year n by applying the de Wit model. Let z be the proportionate yield of Strain 1 and $1 - z$ that of Strain 2, expressed on an equal weight basis. Then

$$z = P_E P_A W \sum_{i=1}^{n-1} S_i x_{n-i}/D \tag{2.25}$$

and

$$1 - z = Q_E Q_A \sum_{i=1}^{n-1} T_i y_{n-i}/D$$

where

$$D = P_E P_A W \sum_{i=1}^{n-1} S_i x_{n-i} + Q_E Q_A \sum_{i=1}^{n-1} T_i y_{n-i}.$$

According to the de Wit model the seed produced by Strain 1 is

$$M_1 k_{12} z / [(k_{12} - 1)z + 1]$$

and that produced by Strain 2 is

$$M_2 k_{21} (1 - z) / [1 - (1 - k_{21})(1 - z)].$$

Discounting these for volunteer species gives the seed produced at the beginning of summer in year n, that is, x_n and y_n. Thus

$$x_n = \frac{P_V M_1 k_{12} z}{(k_{12} - 1)z + 1}, \quad y_n = \frac{P_V M_2 k_{21}(1 - z)}{1 - (1 - k_{21})(1 - z)}. \tag{2.26}$$

These are the basic difference equations for new seed production in Model 2.

We also record here some other important quantities for Model 2. From (2.16) and (2.19), the *total hard seed of all ages* of Strain 1 present in year n is

$$\sum_{i=1}^{n-1} P_S^i H_i x_{n-i} = \sum_{i=1}^{n-1} P_S^i \left\{ \prod_{k=1}^{i} P_k(H \mid B) \prod_{k=1}^{i} P_k(B \mid B) \right.$$

$$+ \prod_{k=1}^{i} P_k(H \mid F) P_1(F \mid B) + \sum_{k=1}^{i-1} P_{k+1}(F \mid B)$$

$$\left. \times \prod_{j=1}^{k} P_j(H \mid B) P_j(B \mid B) \prod_{j=k+1}^{i-1} P_j(H \mid F) \right\} x_{n-i}. \tag{2.27}$$

If we add to this x_n, the new seed production for year n, we obtain the *total seed present* at the beginning of summer in year n, which we call t_n;

$$t_n = x_n + \sum_{i=1}^{n-1} P_S^i H_i x_{n-i}. \tag{2.28}$$

The *proportion of free seed present* in this model in year n is

$$\frac{\sum_{i=1}^{n-1} H_i(F) x_{n-i}}{\sum_{i=1}^{n} H_i x_{n-i}} = \frac{\sum_{i=1}^{n-1} H_i(F) x_{n-i}}{\sum_{i=1}^{n} [H_i(F) + H_i(B)] x_{n-i}} \tag{2.29}$$

where we have not included the current year's production.

There is a submodel of Model 2 which we call Model 2A, in which we assume that burr seed older than one year does not become free. This is an unrealistic assumption but it gives a simpler model than the general model and is a reasonable approximation to our experimental mixtures, as we show later. The quantities $P_i(F \mid B)$, $Q_i(F \mid B)$, $i \geqslant 2$, were not measured in the experiments anyway, but instead "bulked" estimates only of the burr and free seed were available. The equations for Model 2A are easily written down from those of Model 2; we do not give them here.

Model 2A reduces to Model 1 in the case when $P_i(F|B) = 0$, i.e., no free seed whatsoever is assumed. Thus Model 2A is identical to Model 1A if seed of all ages is assumed to soften at the same rate, and no free seed is assumed, since in this case Model 1 is also the same as Model 1A.

"Memory Length" of the Models

In Models 1 and 2 as described so far, there is the potential for seed of any age to soften, i.e. we have made no assumptions on whether the $P_i(S|B)$ or $P_i(S|B)$ are zero or not. In practice we may model a mixture by assuming that only seed up to a certain age N contributes soft seed to the soft seed pool. There is no point in counting seed older than this.

Formally we define N to be the largest integer for which either $P_i(S|F) > 0$ or $P_i(S|B) > 0$, and only count seed up to age N. If the corresponding quantity for Strain 2 differs from N, we take N to be the larger of the two.

However, there is nothing in principle to stop us from taking $N = +\infty$, that is, seed of any age potentially contributes soft seed and indeed we consider this case for some mixtures. (In fact we showed above that Model 1A can be written in this form.)

2.5 Constant soft seed probabilities: approximation to Model 2

In Models 2 and 2A, specification of the probabilities of softening of seed of age 1, 2, 3, ..., N years is required. These individual probabilities were not measured in the experiments reported in Rossiter et al. (1985); instead only a "bulked" softening rate for burr and free seed of all ages was measured. We are led to ask if it is possible to replace the individual softening rates of i-year-old seed with an average figure, without altering the results of the model much. In this section we show that this is possible for Model 2A at least inasmuch as the long-term behavior of the models remains substantially the same. In Chapter 7 we show that the short term behavior is also largely unaltered, at least for the parameter values we use. But our results indicate that the approximation is not good enough for use in the general Model 2.

To define average softening probabilities for Model 2, consider the total amount of hard burr seed in the system in year n. From (2.16) this is proportional to

$$\sum_{i=1}^{n-1} P_S^i \prod_{k=1}^{i-1} P_k(H|B) \prod_{k=1}^{i} P_k(B|B) x_{n-i},$$

while the amount of seed softening from this is, by (2.17),

$$\sum_{i=1}^{n-1} P_S^i P_i(S|B) \prod_{k=1}^{i-1} P_k(H|B) \prod_{k=1}^{i} P_k(B|B) x_{n-i}.$$

By omitting x_{n-i} in these expressions, and replacing $n-1$ by the memory length N, we define an average probability of burr seed softening by

$$P(S|B) = \frac{\displaystyle\sum_{i=1}^{N} P_S^i P_i(S|B) \prod_{k=1}^{i-1} P_k(H|B) \prod_{k=1}^{i} P_k(B|B)}{\displaystyle\sum_{i=1}^{N} P_S^i \prod_{k=1}^{i-1} P_k(H|B) \prod_{k=1}^{i} P_k(B|B)}. \tag{2.30}$$

Similarly from (2.19) and (2.20) we define an average probability of free seed softening by

$$P(S|F) = \frac{\displaystyle\sum_{i=1}^{N} P_S^i P_i(S|F)\left\{\prod_{k=1}^{i-1} P_k(H|F)P_1(F|B) + \sum_{k=1}^{i-1} P_{k+1}(F|B)\prod_{j=1}^{k} P_j(H|B)P_j(B|B)\prod_{j=k+1}^{i-1} P_j(H|F)\right\}}{\displaystyle\sum_{i=1}^{N} P_S^i\left\{\prod_{k=1}^{i-1} P_k(H|F)P_1(F|B) + \sum_{k=1}^{i-1} P_{k+1}(F|B)\prod_{j=1}^{k} P_j(H|B)P_j(B|B)\prod_{j=k+1}^{i-1} P_j(H|F)\right\}}. \tag{2.31}$$

Also let

$$P(H|B) = 1 - P(S|B), \quad P(H|F) = 1 - P(S|F).$$

Consider a new version of Model 2 in which all $P_i(S|B)$ and $P_i(S|F)$, are replaced by $P(S|B)$ and $P(S|F)$, respectively, and the memory length of the model is infinite, that is seed of all ages contributes soft seed at these rates. We now show that the total soft seed produced in this model, a quantity proportional to which determines its long term behavior, is the same as that in the original model, except for a term which is small for large N; *provided* $P_i(F|B) = 0$ for $i \geq 2$, i.e. we are actually in the Model 2A case.

To prove this, let $S_i^{\text{new}}(F)$ be the soft seed produced from i-year-old hard free seed in the new version. Then by (2.20)

$$S_i^{\text{new}}(F) = P_S^i P(S|F)\left\{P^{i-1}(H|F)P_1(F|B)\right.$$

$$\left. + \sum_{k=1}^{i-1} P_{k+1}(F|B)P^k(H|B)\prod_{j=1}^{k} P_j(B|B)P^{i-k-1}(H|F)\right\}$$

so the total soft seed produced in the new system is

$$\sum_{i=1}^{\infty} P_S^i P^{i-1}(H\,|\,F)P_1(F\,|\,B)P(S\,|\,F) + \sum_{k=1}^{\infty}\sum_{i=k+1}^{\infty} P_S^{i-k-1}P^{i-k-1}(H\,|\,F)$$

$$\times\, P_S^{k+1}P_{k+1}(F\,|\,B)P^k(H\,|\,B)\prod_{j=1}^{k} P_j(B\,|\,B)P(S\,|\,F)$$

$$= \frac{P_S P_1(F\,|\,B)P(S\,|\,F)}{1 - P_S P(H\,|\,F)} + \frac{P_S P(S\,|\,F)}{1 - P_S P(H\,|\,F)}$$

$$\times \sum_{k=1}^{\infty} P_S^k P_{k+1}(F\,|\,B)P^k(H\,|\,B)\prod_{j=1}^{k} P_j(B\,|\,B). \tag{2.32}$$

Here we inverted the order of summation in the second sum and used the formula for the sum of an infinite geometric series.

We compare this with the total soft seed produced under the original model, which is

$$\sum_{i=1}^{N} S_i(F) = P(S\,|\,F)\sum_{i=1}^{N} P_S^i \prod_{k=1}^{i-1} P_k(H\,|\,F)P_1(F\,|\,B)$$

$$+ P(S\,|\,F)\sum_{i=1}^{N} P_S^i \sum_{k=1}^{i-1} P_{k+1}(F\,|\,B)$$

$$\times \prod_{j=1}^{k} P_j(H\,|\,B)P_j(B\,|\,B) \prod_{j=k+1}^{i-1} P_j(H\,|\,F)$$

where we substituted the definition (2.31) of $P(S\,|\,F)$ in (2.20).

We have

$$P_S P(H\,|\,F) = P_S[1 - P(S\,|\,F)] = P_S[1 - \text{expression (2.31)}]$$

$$= P_S \frac{\sum_{i=1}^{N} P_S^i P_i(H\,|\,F)\left\{\prod_{k=1}^{i-1} P_k(H\,|\,F)P_1(F\,|\,B) + \prod_{k=1}^{i-1} P_{k+1}(F\,|\,B)\prod_{j=1}^{k} P_j(H\,|\,B)P_j(B\,|\,B)\prod_{j=k+1}^{i-1} P_j(H\,|\,F)\right\}}{\sum_{i=1}^{N} P_S^i \left\{\prod_{k=1}^{i-1} P_k(H\,|\,F)P_1(F\,|\,B) + \prod_{k=1}^{i-1} P_{k+1}(F\,|\,B)\prod_{j=1}^{k} P_j(H\,|\,B)P_j(B\,|\,B)\prod_{j=k+1}^{i-1} P_j(H\,|\,F)\right\}}.$$

$$\tag{2.33}$$

Subtracting this expression from unity we obtain an expression whose numerator is

$$\left\{\sum_{i=1}^{N} P_S^{i-1}\prod_{k=1}^{i-1} P_k(H\,|\,F) - \sum_{i=1}^{N} P_S^i P_i(H\,|\,F)\prod_{k=1}^{i-1} P_k(H\,|\,F)\right\}P_1(F\,|\,B)P_S$$

$$+ \sum_{i=1}^{N} P_S^i \sum_{k=1}^{i-1} P_{k+1}(F\,|\,B)\prod_{j=1}^{k} P_j(H\,|\,B)P_j(B\,|\,B)\prod_{j=k+1}^{i-1} P_j(H\,|\,F)$$

$$- \sum_{i=1}^{N} P_S^{i+1}P_i(H\,|\,F)\sum_{k=1}^{i-1} P_{k+1}(F\,|\,B)\prod_{j=1}^{k} P_j(H\,|\,B)P_j(B\,|\,B)\prod_{j=k+1}^{i-1} P_j(H\,|\,F)$$

$$= \left\{ \sum_{i=1}^{N} P_S^{i-1} \prod_{k=1}^{i-1} P_k(H\,|\,F) - \sum_{i=1}^{N} P_S \prod_{k=1}^{i} P_k(H\,|\,F) \right\} P_1(F\,|\,B) P_S$$

$$+ \sum_{i=1}^{N} P_S^{i} \sum_{k=1}^{i-1} P_{k+1}(F\,|\,B) \prod_{j=1}^{k} P_j(H\,|\,B) P_j(B\,|\,B) \prod_{j=k+1}^{i-1} P_j(H\,|\,F)$$

$$- \sum_{i=1}^{N} P_S^{i+1} \sum_{k=1}^{i} P_{k+1}(F\,|\,B) \prod_{j=1}^{k} P_j(H\,|\,B) P_j(B\,|\,B) \prod_{j=k+1}^{i} P_j(H\,|\,F)$$

$$+ \sum_{i=1}^{N} P_S^{i+1} P_{i+1}(F\,|\,B) \prod_{j=1}^{i} P_j(H\,|\,B) P_j(B\,|\,B)$$

$$= \left[1 - P_S^{N} \prod_{k=1}^{N} P_k(H\,|\,F) \right] P_1(F\,|\,B) P_S$$

$$- P_S^{N+1} \sum_{k=1}^{N} P_{k+1}(F\,|\,B) \prod_{j=1}^{k} P_j(H\,|\,B) P_j(B\,|\,B) \prod_{j=k+1}^{N} P_j(H\,|\,F)$$

$$+ \sum_{i=1}^{N} P_S^{i+1} P_{i+1}(F\,|\,B) \prod_{j=1}^{i} P_j(H\,|\,B) P_j(B\,|\,B).$$

Since the denominator of (2.33) is $\Sigma_{i=1}^{N} S_i(F)/P_S P(S\,|\,F)$, we have

$$\frac{P_S P_1(F\,|\,B) P(S\,|\,F)}{1 - P_S P(H\,|\,F)} = \sum_{i=1}^{N} S_i(F) \left\{ 1 - P_S^{N} \prod_{k=1}^{N} P_k(H\,|\,F) \right.$$

$$- P_S^{N} \sum_{k=1}^{N} P_{k+1}(F\,|\,B) \prod_{j=1}^{k} P_j(H\,|\,B) P_j(B\,|\,B)$$

$$\times \prod_{j=k+1}^{N} P_j(H\,|\,F)/P_1(F\,|\,B) + \sum_{k=1}^{N} P_S^{k} P_{k+1}(F\,|\,B)$$

$$\left. \times \prod_{j=1}^{k} P_j(H\,|\,B) P_j(B\,|\,B)/P_1(F\,|\,B) \right\}^{-1}$$

$$= \sum_{i=1}^{N} S_i(F) \left\{ 1 - P_S^{N} \prod_{k=1}^{N} \dot{P}_k(H\,|\,F) - E_N \right\}^{-1},$$

say. Hence from (2.32)

$$\sum_{i=1}^{\infty} S_i^{\text{new}}(F) = \sum_{i=1}^{N} S_i(F) \left\{ 1 - P_S^{N} \prod_{k=1}^{N} P_k(H\,|\,F) - E_N \right\}^{-1}$$

$$+ \frac{P_S P(S\,|\,F)}{1 - P_S P(H\,|\,F)} \sum_{i=1}^{\infty} P_S^{i} P_{i+1}(F\,|\,B) P^{i}(H\,|\,B) \prod_{j=1}^{i} P_j(B\,|\,B)$$

$$= \sum_{i=1}^{N} S_i(F) \left[1 - P_S^{N} \prod_{k=1}^{N} P_k(H\,|\,F) - E_N \right]^{-1}$$

$$\times \left\{ 1 + \sum_{i=1}^{\infty} P_S^{i} P_{i+1}(F\,|\,B) P^{i}(H\,|\,B) \prod_{j=1}^{i} P_j(B\,|\,B)/P_1(F\,|\,B) \right\}.$$

A similar, but easier analysis shows that

$$\sum_{i=1}^{\infty} S_i^{\text{new}}(B) = \sum_{i=1}^{N} S_i(B)\left[1 - P_S^N \prod_{k=1}^{N} P_k(H\,|\,B)\right]^{-1}.$$

Now suppose we are in Model 2A, that is, $P_{i+1}(F\,|\,B) = 0$ for $i \geqslant 1$. Since $E_N = 0$ in this case the above shows that $\Sigma_{i=1}^{\infty} S_i^{\text{new}}$ and $\Sigma_{i=1}^{N} S_i$ differ only by terms of order $P_S^N \Pi_{k=1}^{N} \max\{P_k(H\,|\,B), P_k(H\,|\,F)\}$; this is small even for moderate values of N and vanishes as $N \to \infty$. This means that the long term behavior of the models is similar for moderate values of N and identical as $N \to \infty$. But in general, E_N and $\Sigma_{i=1}^{\infty} P_S^i P_{i+1}(F\,|\,B) P^i(H\,|\,B) \Pi_{j=1}^{i} P_j(B\,|\,B)/P_1(F\,|\,B)$ need not be small, so in the most general case of Model 2, we cannot replace the age dependent softening probabilities with the average softening probabilities defined by (2.30) and (2.31).

Incidentally, we tried to use the above procedure to define an "average" probability of burr seed becoming free in Model 2 or 2A; but a similar procedure does not lead to models with the same long term behavior.

3

Analysis of model G: Introduction

3.1   Introduction

Our aim in this and the following three chapters is to gain an understanding of the large time, or asymptotic, behavior of the sequence (t_n, u_n) describing the maximum seed pools for each strain. The asymptotic behavior of the residual hard seed numbers and so forth, can then be inferred from that of the maximum seed numbers. The analysis of (t_n, u_n) takes place via the difference equations which were derived in Chapter 2. In this Chapter, and the next two, we consider only Model 1A, which is recast in a more general and more convenient form as Model G. Chapter 6 considers the more general models.

We include in Section 2 of this chapter an elementary introduction to the stability concepts we require. Section 3 is concerned with preliminary results on Model G. Chapters 4 and 5 will continue this investigation with a deeper analysis of Model G. We conclude this chapter with a review of a variety of topics related to the content of the work described in this monograph, that is, difference equation models of annual plant populations having seed banks.

We introduce the following notation which is defined in terms of that established in Chapter 2. Let

$$U = M_1 P_V P_E P_A W P(S) P_S \qquad V = M_2 P_V Q_E Q_A Q(S) Q_S$$

$$L = M_1 P_V \qquad\qquad M = M_2 P_V$$

$$\alpha = P_S P(H) \qquad\qquad \beta = Q_S Q(H)$$

$$X = U/L \qquad\qquad Y = V/M.$$

With this notation, the equations governing the year to year maximum seed pools in Model 1A can be rewritten from (2.13) as

$$t_{n+1} = \frac{k_{12} U t_n}{k_{12} X t_n + Y u_n} + \alpha t_n, \quad u_{n+1} = \frac{k_{21} V u_n}{X t_n + k_{21} Y u_n} + \beta u_n. \quad (3.1)$$

We can simplify and generalise the above system by defining

$$t_{n+1} = LF(z_n) + \alpha t_n \tag{3.2a}$$

$$u_{n+1} = MG(1 - z_n) + \beta u_n \tag{3.2b}$$

where

$$z_n = \frac{Xt_n}{Xt_n + Yu_n}$$

and the functions F and G are defined, continuous and strictly increasing on [0, 1], differentiable on (0, 1), and

$$F(0) = G(0) = 0 \quad \text{and} \quad F(1) = G(1) = 1.$$

The system (3.1) is then the special case given by choosing

$$F(z) = \frac{k_{12}z}{(k_{12} - 1)z + 1} \quad \text{and} \quad G(z) = \frac{k_{21}z}{(k_{21} - 1)z + 1}. \tag{3.3}$$

Note that when $k_{12}k_{21} = 1$ each of these functions is the inverse of the other.

Recall from Chapter 1 that P_V may, and usually will, be a function of n and decreases to a *positive* limit as $n \to \infty$. This has the effect of rendering the system (3.2) *non-autonomous*, meaning that time enters otherwise than through the u_ns and t_ns. In fact time enters through the factors L and M – the factor P_V does not enter X or Y. In practice this time variation is not a problem because, within the limits of accuracy of field observations, P_V actually attains its limiting value after a few seasons, typically by the seventh or eighth. Thus time variation of P_V ceases after n reaches a value $n(V)$, say, a small positive integer. Consequently Equation (3.2) can be used to calculate $t_{n(V)}$ and $u_{n(V)}$, and then its subsequent behavior will be as described below in Chapters 4 and 5 under the assumption that P_V is stationary in time and with the initial values $t_{n(V)}$ and $u_{n(V)}$. Hence we lose no generality of application by treating (3.2) as if P_V is independent of n for $n \geq 1$, and we assume this in the sequel unless stated otherwise. It is worth observing that a thorough treatment of the case of time varying P_V should take explicit account of the grass component of the pasture and, in particular, its competitive interaction with the clover components. We will not address that problem in this monograph.

The asymptotic analysis of (3.2) is facilitated by rescaling the variables to eliminate the constants L and M. Define p_n and q_n by

$$t_n = \frac{L}{1 - \alpha} p_n \quad \text{and} \quad u_n = \frac{U}{1 - \alpha} \frac{q_n}{Y} \tag{3.4}$$

and let

$$\rho = \frac{V(1-\alpha)}{U(1-\beta)}.$$ (3.5)

Equations (3.2a) and (3.2b) can be recast as

$$p_{n+1} = (1-\alpha)F\left(\frac{p_n}{p_n + q_n}\right) + \alpha p_n$$ (3.6a)

$$q_{n+1} = \rho(1-\beta)G\left(\frac{q_n}{p_n + q_n}\right) + \beta q_n$$ (3.6b)

and it is this system, called *Model G*, which we discuss in this and the following two chapters, together with the following particular form arising from Model 1A,

$$p_{n+1} = (1-\alpha)\frac{k_{12}p_n}{k_{12}p_n + q_n} + \alpha q_n$$ (3.7a)

$$q_{n+1} = (1-\beta)\rho\frac{k_{21}q_n}{p_n + k_{21}q_n} + \beta q_n.$$ (3.7b)

We will always assume $0 < \alpha, \beta < 1$.

Equations (3.6a) and (3.6b) determine a two dimensional *dynamical system* and since the results we obtain are best couched in the terminology of dynamical system theory, we shall present the rudiments of this theory in the next section. Before doing so, however, we complete this introduction with a few general remarks which will be relevant when age dependent models are discussed in Chapter 6.

By (2.13) and the notation introduced above, the new seed production (x_n, y_n) satisfies $x_n = t_n - \alpha t_{n-1}$, $y_n = u_n - \beta u_{n-1}$. Then (3.2) can be used to express x_n and y_n in terms of z_{n-1}, and z_{n-1} can be eliminated to give the following relationship between the yearly seed production numbers of the two species,

$$y_n/M = G(1 - F^{-1}(x_n/L)).$$ (3.8)

This relationship depends only on the mechanisms controlling the interactions of the two species at the time of seed production. It will provide the key to understanding the age-dependent case when F and G are given by (3.3), and in this case (3.8) reduces to

$$y_n/M = \frac{k_{12}k_{21}(1 - x_n/L)}{k_{12}k_{21}(1 - x_n/L) + x_n/L}.$$ (3.9)

Note that when $k_{12}k_{21} = 1$ this is a simple linear relationship,

$$x_n/L + y_n/M = 1.$$ (3.10)

If the last equation is re-expressed in terms of t_ns and u_ns, using (3.2b), the resulting difference equation can be solved in the following form,

$$u_{n+1} = k\beta^n + M\frac{1-\beta^n}{1-\beta} - \frac{M}{L}\left[\sum_{i=0}^{n}\beta^{n-i}t_{i+1} - \alpha\sum_{i=0}^{n-1}\beta^{n-1-i}t_{i+1}\right],$$

representing the sequence (u_n) as a moving average of (t_n). When P_V is constant, this expression can be used to convert the autonomous system (3.1) into a non-autonomous difference equation for (t_n). This difference equation seems to be intractable.

3.2   Rudiments of dynamical system theory

Let $\mathbb{R}_+$ denote the set of non-negative real numbers and $\mathscr{S} = \mathbb{R}_+^m$ be the set of m-vectors having non-negative components, where m is a positive integer. Elements of $\mathscr{S}$ will be denoted by bold face letters, for example $\mathbf{x}_n$ or $\boldsymbol{\varepsilon}$. For the models discussed in the sequel we will have $m = 2$ or $m = 2N$ where N is the memory length defined in Chapter 2 in connection with Model 2. To describe (3.6), for example, we take $m = 2$ and $\mathbf{x}_n = (p_n, q_n)$.

Let $\mathscr{T} : \mathscr{S} \to \mathscr{S}$ be a continuous mapping of $\mathscr{S}$ into itself. This mapping defines a discrete (semi) dynamical system with state space $\mathscr{S}$ as follows. Let $\mathbf{x}_1 \in \mathscr{S}$ and working inductively define $\mathbf{x}_{n+1} = \mathscr{T}(\mathbf{x}_n)$ for $n = 1, 2, \ldots$. The sequence $\{\mathbf{x}_n\}$ is called *the orbit starting from* $\mathbf{x}_1$, and if we do not wish to fix attention on the initial point, it is simply called an orbit. We can write the transformation which takes $\mathbf{x}_1$ into $\mathbf{x}_{n+1}$ as $\mathbf{x}_{n+1} = \mathscr{T}_n(\mathbf{x}_1)$ where $\mathscr{T}_n$ denotes the n-fold iterate of $\mathscr{T}$, and it is conventional to define $\mathscr{T}_0$ to be the identity mapping on $\mathscr{S}$.

Denote the set of natural numbers $\{1, 2, \ldots\}$ by $\mathbb{N}$. Strictly speaking, the term "dynamical system" applies to the mapping $\pi : \mathbb{N} \times \mathscr{S} \to \mathscr{S}$ whose values are given by $\pi(n, \mathbf{x}) = \mathscr{T}_{n-1}(\mathbf{x})$. For our purposes it is sufficient to think of the collection of all orbits, or $\mathscr{T}$, as constituting the dynamical system. If $\mathscr{T}$ has a well-defined inverse then two orbits which start from different points either have no points in common or one is contained in the other. The corresponding dynamical system is then called *invertible*.

The aspect of a dynamical system which will be of most interest to us is its asymptotic behaviour. Suppose a point $\boldsymbol{\varepsilon}$ satisfies the equation $\mathbf{x} = \mathscr{T}(\mathbf{x})$. Any orbit starting from $\boldsymbol{\varepsilon}$ consists only of this point and accordingly we call $\boldsymbol{\varepsilon}$ an *equilibrium point*. If at least one component of $\boldsymbol{\varepsilon}$ is zero, we call $\boldsymbol{\varepsilon}$ a *boundary equilibrium* and otherwise we call $\boldsymbol{\varepsilon}$ an *interior equilibrium*. The dynamical systems we shall investigate always have boundary equilibria and for some parameter values they will also have interior equilibria. In the context of population modeling the attainment of a boundary equilibrium corresponds to a steady state situation in which some species are absent and the attainment of an interior equilibrium corresponds to a state of perpetual co-existence of all species.

Since an equilibrium represents a steady state it is of some interest to understand the behavior of orbits which start near an equilibrium point. If such points never wander too far from the equilibrium point we say that it is *stable*. More specifically ε is stable if given a neighborhood U of ε, there is another neighborhood W contained in U such that whenever $\mathbf{x} \in U$ then $\mathcal{T}_n(\mathbf{x}) \in W$ for all $n \in \mathbb{N}$. If ε is not stable it is called *unstable*.

Call ε an *attracting point* if there is a neighborhood U of ε such that $\lim_{n \to \infty} \mathcal{T}_n(\mathbf{x}) = \varepsilon$ for all $\mathbf{x} \in U$. Since the limit may not exist, it is fairly obvious that ε can be stable without being attracting, but perhaps it is less clear that ε can be attracting without being stable. This can occur if the orbit swings away from ε before converging on it and the magnitude of this first excursion increases unboundedly as the initial point is taken closer and closer to ε. To put it another way, ε is attracting but unstable if $\mathcal{T}_n(\mathbf{x}) \to \varepsilon$ as $n \to \infty$ non-uniformly with respect to some infinite set of initial points $\mathbf{x}$ tending to ε. Frequently, however, both attraction and stability occur together and ε is then called *asymptotically stable*, abbreviated as AS.

Asymptotic stability is relatively easy to detect. Let $\mathcal{J}(\varepsilon)$ be the Jacobian matrix of $\mathcal{T}$ evaluated at ε, where, if ε is a boundary equilibrium, some of the derivatives defining $\mathcal{J}(\varepsilon)$ are evaluated as one sided limits. Let $r(\varepsilon)$ be the spectral radius of $\mathcal{J}(\varepsilon)$, that is, the magnitude of the eigenvalue of $\mathcal{J}(\varepsilon)$ having the largest modulus. The so-called *linearization theorem* states that ε is asymptotically stable if $r(\varepsilon) < 1$; see LaSalle (1977). When $r(\varepsilon) = 1$, no decision can be made and a finer analysis is necessary. Lessard and Karlin (1982) give some results in this direction. When $r(\varepsilon) = 1$ then ε is called *neutrally stable*, a term which really describes $r(\varepsilon)$ rather than the behaviour of orbits starting near ε.

Let $\mathcal{S}^0$ denote the interior of $\mathcal{S}$, that is $\mathcal{S}^0$ is the set of m-vectors with positive components. If $\lim_{n \to \infty} \mathcal{T}_n(\mathbf{x}) = \varepsilon$ for all $\mathbf{x} \in \mathcal{S}^0$ we call ε a globally attracting point. If ε is also stable, it is called *globally asymptotically stable* (GAS). There do exist some general techniques for establishing that a point is GAS, but their implementation can be very difficult in individual cases.

For example, if $m = 2$ it may suffice to find a subset $\mathcal{R} \subset \mathcal{S}$ which contains ε and is (*positively*) *invariant*, by which we mean $\mathcal{T}(\mathbf{x}) \in \mathcal{R}$ for each $\mathbf{x} \in \mathcal{R}$. This technique will be applied in the next two chapters to the dynamical system defined by (3.6).

A second technique is in principle universally applicable. Let $V: \mathscr{S} \to \mathbb{R}$ be a continuous real valued function. Define the function $\dot{V}(\mathbf{x}) = V(\mathcal{T}(\mathbf{x})) - V(\mathbf{x})$. If $\dot{V}(\mathbf{x}) \leqslant 0$ for each $\mathbf{x} \in \mathscr{S}^0$ then V is called a *Liapunov* function. Suppose also that

(i) $V(\mathbf{x}) > 0$ if $\mathbf{x} \neq \boldsymbol{\varepsilon}$, but $V(\boldsymbol{\varepsilon}) = 0$;
(ii) $V(\mathbf{x}) \to \infty$ as $\| \mathbf{x} \| \to \infty$; and
(iii) $\dot{V}(\mathbf{x}) < 0$ if $\mathbf{x} \neq \boldsymbol{\varepsilon}$.

Then $\boldsymbol{\varepsilon}$ is GAS.

Establishing that $\boldsymbol{\varepsilon}$ is GAS may be carried out in three stages. Firstly the linearization theorem may be used to establish that $\boldsymbol{\varepsilon}$ is AS for a range of parameter values. Secondly, a prima facie case that $\boldsymbol{\varepsilon}$ is GAS my be built by visual inspection of some computed orbits, and the third stage is the actual *proof* that $\boldsymbol{\varepsilon}$ is GAS, say, by finding a suitable Liapunov function. In specific cases the last stage may present intractable problems since there is no universal recipe for constructing Liapunov functions. In modeling population processes it is usual to omit stage three and rely on stage one, and perhaps also on stage two. A significant feature of our work is that the equilibria of the system (3.6) can be completely classified in all relevant cases. In particular, in Chapter 5 we shall exhibit a Liapunov function for the system (3.7) when it has an AS interior equilibrium.

A very readable survey of dynamical system theory which covers all that we shall need is that of LaSalle (1977). Whitely (1983) concentrates on one and two dimensional systems and discusses phenomena which will not occur with our well behaved systems. His paper is a useful introduction to the general theory of dynamical systems. Finally, Goh (1980) gives a good account with many examples of the use of Liapunov functions.

3.3 Some preliminary results for Model G

In this section we prove some preliminary results which require no conditions beyond those already imposed in defining Model G. These will serve as an introduction to the more detailed results of Chapters 4 and 5.

Theorem 3.1
The dynamical system defined by (3.6) is invertible.

Proof. Denote the image under the transformation (3.6) of an arbitrary point (p, q) in $\mathscr{S}^0$ ($m = 2$) by (p', q'). It is obvious that $(p', q') \in \mathscr{S}^0$, and we must show that (p, q) is uniquely determined by (p', q'). Let

$r = q/p$. Elimination of the βq term in (3.6b) by using (3.6a) yields

$$q' - (1 - \beta)\rho G\left(\frac{r}{1+r}\right) = \alpha^{-1}\beta r\left(p' - (1 - \alpha)F\left(\frac{1}{1+r}\right)\right).$$

(3.11)

The left hand side of this equation is a strictly decreasing function of r, decreasing from q' to $q' - (1 - \beta)\rho$ as r increases from zero to infinity. Similarly $p' - (1 - \alpha)F((1 + r)^{-1})$ increases from $p' - (1 - \alpha)$ to p'. When $p' < 1 - \alpha$ denote by $\tilde{r}$ the unique positive solution of $p' = (1 - \alpha)F((1 + r)^{-1})$, and when $p' \geq 1 - \alpha$ let $\tilde{r} = 0$. Then the right hand side of (3.11) is strictly increasing from zero to ∞ in the interval $[\tilde{r}, \infty)$. It follows that (3.11) has exactly one solution for r in $(\tilde{r}, \infty)$. Having determined a unique r corresponding to (p', q'), the unique preimage (p, q) can be calculated from (3.6).

Our second result shows that a weak form of stability always obtains.

Theorem 3.2

The dynamical system defined by (3.6) is Lagrange stable, that is, the sequences (p_n) and (q_n) are bounded.

Proof. Since $G(z) \leq 1$, Equation (3.6b) yields

$$q_{n+1} \leq \rho(1 - \beta) + \beta q_n \leq \ldots \leq \rho(1 - \beta^n) + \beta^n q_1$$

whence $\overline{\lim}_{n\to\infty} q_n \leq \rho$. Similarly $\overline{\lim}_{n\to\infty} p_n \leq 1$.

Remark. It follows that all seed yields are bounded and, moreover, if $P_V \to 0$ as $n \to \infty$ then the seed numbers tend to zero.

The next result deals with the question of asymptotic stability of the boundary equilibria of the dynamical system (3.6).

Theorem 3.3

(i) *The system (3.6) has boundary equilibria at $(1, 0)$ and $(0, \rho)$. Suppose the one-sided derivatives $F'(0+)$, $G'(0+)$, $F'(1-)$ and $G'(1-)$ are finite.*

(ii) *If*

$$\rho G'(0+) < 1$$

(3.12)

then $(1, 0)$ is AS, and it is unstable if the inequality (3.12) is reversed.

(iii) *If*

$$F'(0+) < \rho$$

(3.13)

then $(0, \rho)$ is AS and it is unstable if the inequality (3.13) is reversed.

Proof. It is trivial to check that the given points are equilibria. If $x = (p, q) \in \mathscr{S}$ then the Jacobian at x of the transformation defined by (3.6) is

$$\mathscr{J}(x) = \begin{bmatrix} \alpha + \dfrac{(1-\alpha)q}{(p+q)^2}F'(z) & -\dfrac{(1-\alpha)p}{(p+q)^2}F'(z) \\[4mm] -\dfrac{(1-\beta)\rho q}{(p+q)^2}G'(1-z) & \beta + \dfrac{(1-\beta)\rho p}{(p+q)^2}G'(1-z) \end{bmatrix} \tag{3.14}$$

It is easily checked that the eigenvalues of $\mathscr{J}(1, 0)$ are α and $\beta + (1-\beta)\rho G'(0+)$, and those of $\mathscr{J}(0, \rho)$ are β and $\alpha + (1-\alpha)\rho^{-1}F'(0+)$. The assertions about stability now follow from the linearization theorem. This completes the proof of Theorem 3.3.

We now consider the existence of interior equilibria. These coincide with intersections of the *isoclines* of the system (3.6). The curve on which the increment $p' - p$ is zero is called the p-isocline and denoted by $\mathscr{I}_F$. It is given by solving the following equation for q:

$$0 = (1-\alpha)F\left(\frac{p}{p+q}\right) + \alpha p - p,$$

that is,

$$p = F\left(\frac{p}{p+q}\right). \tag{3.15}$$

This has a unique solution for q expressible as

$$q = i(p) = p(1/\phi(p) - 1) \tag{3.16}$$

where $\phi(\cdot)$ is the inverse of $F(\cdot)$, and, from (3.15), the function $i(\cdot)$ is defined on [0, 1]. Similarly the q-isocline, $\mathscr{I}_G$, is the curve on which $q' - q = 0$ and it is given by

$$q = \rho G\left(\frac{q}{p+q}\right) \tag{3.17}$$

which can be solved for p in the form

$$p = \rho j(q/\rho) \quad (0 \leqslant q \leqslant \rho) \tag{3.18}$$

where

$$j(q) = q(1/\gamma(q) - 1) \quad (0 \leqslant q \leqslant 1)$$

and $\gamma(\cdot)$ is the inverse function of $G(\cdot)$.

The isoclines are smooth curves, indeed, the functions defined by (3.16) and (3.18) are differentiable. It follows from (3.15) that $q \to 0+$

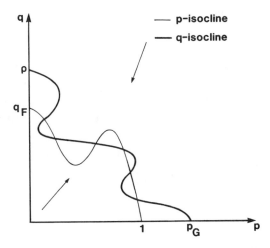

Figure 3.1. Schematic illustration of $\mathscr{I}_F$, $\mathscr{I}_G$ for Model G.

as $p \to 1-$, that is, $i(1-) = 0$. Again, from (3.16),

$$\lim_{p \to 0+} q = i(0+) = 1/\phi'(0+) = F'(0+).$$

We shall always assume that $0 < F'(0+) < \infty$ and in that case $\mathscr{I}_F$ is a smooth curve joining the points $(0, q_F)$ and $(1, 0)$ where

$$q_F = F'(0+).$$

Similarly, always assuming $0 < G'(0+) < \infty$, $\mathscr{I}_G$ is a smooth curve joining the point $(0, \rho)$ to $(p_G, 0)$ where

$$p_G = \rho G'(0+).$$

Figure 3.1 illustrates these concepts. It is important to keep in mind that $\mathscr{I}_G$ is specified by a function of q, and not of p, and hence when viewed as in Figure 3.1 its graph may give the appearance of corresponding to a "multivalued" function. A related fact is that the isoclines need not be monotonic. It is not difficult to show that $\mathscr{I}_F$ is non-increasing if

$$(1 - z)F'(z) \leqslant z^{-1}F(z) \quad (0 < z < 1)$$

and when F is twice differentiable in $(0, 1)$, $\mathscr{I}_F$ is convex downwards (respectively, upwards) if

$$zF''(z) \geqslant (\text{respectively}, \leqslant)2F'(z)(zF'(z) - F(z)) \quad (0 < z < 1).$$

Similar conditions can be obtained in relation to $\mathscr{I}_G$.

If the point $(p, q) \in \mathscr{S}^0$ lies on *both* isoclines then $p' = p$ and $q' = q$, that is, (p, q) is an internal equilibrium. Putting it another way, (p, q) is an internal equilibrium if p and q are positive and satisfy both (3.15) and (3.17). There may be, as suggested by Figure 3.1, many internal equilibria. We will impose conditions that ensure only two cases occur, namely, *either* there are no internal equilibria *or* there is exactly one internal equilibrium. These cases, which will be treated in Chapters 4 and 5 respectively, are the only possibilities for Model 1A. The former case occurs when $k_{12}k_{21} = 1$ and the latter may occur only if $k_{12}k_{21} \neq 1$.

There are obviously no internal equilibria if either $\mathscr{I}_F$ lies completely above $\mathscr{I}_G$ in $\mathscr{S}^0$, or vice versa. When the points of intersection of the isoclines with the co-ordinate axes are distinct, then necessary conditions for the absence of internal equilibria are

$$1 < p_G \quad \text{and} \quad q_F < \rho \tag{3.19a}$$

or

$$p_G < 1 \quad \text{and} \quad \rho < q_F. \tag{3.19b}$$

Theorem 3.3 says that in either case exactly one of the boundary equilibria is AS, and it is then a reasonable conjecture that the AS equilibrium is GAS. We shall see in the next chapter that this indeed is the case. In fact we will be able to build up a quite detailed picture of the behavior of the system's orbits.

The simple argument above shows that necessary conditions for the existence of a single internal equilibrium are

$$1 < p_G \quad \text{and} \quad \rho < q_F \tag{3.20a}$$

or

$$p_G < 1 \quad \text{and} \quad q_F < \rho. \tag{3.20b}$$

Invoking Theorem 3.3 once again, we see that in the first case both boundary equilibria are unstable, and in the second case they are both AS. Thus one might conjecture that (3.20a) ensures that orbits are bounded away from the boundary of $\mathscr{S}$, $\partial \mathscr{S}$, that is, there is some form of coexistence. We shall see in Chapter 5 that the system behaves very smoothly; indeed the internal equilibrium is GAS. On the other hand, when (3.20b) holds the internal equilibrium is not AS and orbits converge to one or other of the boundary equilibria. The initial point of an orbit determines which boundary equilibrium it approaches.

3.4 Review of related work

The system defined by (3.7) seems to be new to population mathematics, although there is some work in the literature on aspects of systems somewhat similar to (3.7). In addition, certain theoretical aspects of the dynamics of annual plant populations having a seed bank have been examined in the research literature. It is our intention in this section to briefly review the main contributions of this literature and hence provide a context for the work described in later chapters. Existing research falls into one of three categories. First, there are investigations of general systems of difference equations, of which (3.6) is a representative. Secondly, considerable attention has been given to models of the evolution of seed dormancy and, finally, some attention has been given to modeling the growth dynamics of annual plant populations taking account of botanically important factors, including dormancy. We new review each of these categories in turn. The unified notation used in this section means that there will be apparent deviations from that used in the work we review.

Consider the general two dimensional system defined by

$$p_{n+1} = p_n f(p_n, q_n), \quad q_{n+1} = q_n g(p_n, q_n) \tag{3.21}$$

where f and g are positive functions defined on $\mathscr{S}$. Analysis of such a general system requires the imposition of regularity conditions on the determining functions f and g, but such conditions need to be chosen to provide a reasonable compromise between generality and a detailed specification of some ecologically interesting situation. Hutson and Moran (1982) and Schumacher (1983) take (3.21) to be a description of predator-prey interactions and they impose conditions on f and g which are consistent with that interpretation. Their conditions ensure that the system is *permanent*, that is, there is a compact subset $\mathscr{D}$ of $\mathscr{S}^0$ which has the following properties:

(i) $\mathscr{D}$ is invariant; and
(ii) an orbit starting in $\mathscr{S} \backslash \mathscr{D}$ eventually enters $\mathscr{D}$.

This ensures continued co-existence of prey and predator because the sequences $\{p_n\}$ and $\{q_n\}$ are bounded away from zero and infinity. However, it could be that orbits behave very erratically within $\mathscr{D}$. The sets of conditions used in these memoirs are quite different although they have overlapping domains of applicability. Schumacher gives additional conditions ensuring that orbits oscillate around a single unstable equilibrium point.

The concepts in these papers have been applied to particular cases of (3.21). For example Hofbauer, Hutson, and Jansen (1987) obtain many results about the permanence of a discrete-time version of the Lotka-Volterra equations. In two dimensions their model specifies that $f(p, q) = \exp[r - ap - bq]$, where r, a and b are constants, and $g(p, q)$ has a similar form. The special case of symmetric competition in which $a = r$ and $g(p, q) = f(q, p)$ has been given detailed consideration by

Jiang and Rogers (1987). They show that this system is permanent but that the orbits can behave in very complicated ways within the set $\mathscr{D}$.

The system (3.21) was considered earlier by Shapiro (1974) who imposed conditions on f and g which reflect competitive interactions between two species. He calls the system (3.21) *remote* from the p-axis if there is a closed set $\mathscr{D}$ satisfying (i) and (ii) above and $\mathscr{D} \cap \{(p, 0): p > 0\} = \phi$, and *drawn* or *attracted* to the p-axis if $q_n \to 0$ as $n \to \infty$ for any $(p_1, q_1) \in \mathscr{S}^0$. Remoteness and attraction to the q-axis are similarly defined. Shapiro stated, without proof, some conditions relating to remoteness or attraction of the axes. Most closely related to our work is an incompletely specified "extension" of Theorem 3.3 above. Shapiro states that if "f and g increase monotonically or satisfy other supplemental conditions," then the following are true:

(a) If (3.20a) holds then the system (3.6) is remote from both axes;

(b) If (3.20b) holds then (3.6) is drawn toward both axes;

(c) If (3.19a) holds then (3.6) is remote from the p-axis and is drawn to the q-axis.

An obvious analogue of (c) holds for (3.19b).

The system (3.7) violates some of the conditions imposed in each of the cited references. These violations include unboundedness of f and g near the origin, and monotonicity properties opposite to those specified. However, the additional structure of (3.6) allows us to derive stronger versions of Shapiro's assertions.

In a paper which seems to be an abstract of a longer unpublished memoir, Szlenk and Zelawski (1985) have propounded a particular version of (3.21) to model the day-by-day growth of a plant. The dry mass M_n of a plant on day n has components W_n and V_n, which are the dry weights of the assimilatory and non-assimilatory parts, respectively, of the plant. Let $\lambda_n = W_n/V_n$. Their model is

$$W_{n+1} = W_n + \beta(\lambda_n, \delta)\alpha(W_n)W_n$$

and

$$V_{n+1} = V_n + [1 - \beta(\lambda_n, \delta)]\alpha(W_n)W_n,$$

where $\alpha(W_n)W_n$ is the increase of total dry weight on day n produced by the assimilatory part of the plant and $\beta(\lambda_n, \delta)$ is the proportion of that increase allocated to the assimilatory part of the plant. This allocation function is tuned by the parameter δ which represents the influence of environmental conditions. The functions $\alpha(.)$ and $\beta(., \delta)$ are assumed to be continuous and decreasing, $\alpha(0) > 0$, $\beta(0, \delta) = 1$ and

$\alpha(\infty) = \beta(\infty, \delta) = 0$. Szlenk and Zelawski analyse the system above using the map which takes (M_n, λ_n) into (M_{n+1}, λ_{n+1}). They obtain properties quite unlike those we obtain for (3.6). For example $M_n \to \infty$ at a rate that is slower than any increasing geometric sequence.

Li (1988) considers the special case of (3.21) where $f(p, q) = (q/p) \times (\alpha - f(q))$ and $g(p, q) = (p/q)(\beta - g(q))^+$, α and β are positive constants, and f and g are strictly increasing positive functions satisfying $f(0) = g(0) = 0$. This model represents the season-to-season fluctuations in total biomass of the juvenile and adult forms in a population of a given organism. All juveniles born at time n survive to adulthood at $n + 1$, and these adults reproduce at time $n + 2$, and then die. The juvenile biomass produced by unit amount of adult biomass is denoted by α, and β denotes the rate of biomass production caused by the juvenile to adult phase transition. Biomass conversion is limited by increased adult densities, and this effect is described by the functions f and g. The focus of interest is the determination of conditions which ensure various categories of persistence. For example, the author defines strong persistence to mean that juvenile and adult biomasses are always positive and that their upper limit points, as $n \to \infty$, are positive. This is a weaker notion than permanence. If $\alpha\beta \leqslant 1$ then p_n, $q_n \to 0$, and if the converse inequality holds then, subject to technical conditions involving f, g and the initial conditions, the system is strongly persistent.

Over the last twenty years there have been several studies of the evolution of seed dormancy in annual plants. The essential idea in these studies is that because of temporal fluctuations in environmental quality, annual plant species may survive better if they can disperse their seed in time. Thus it may be advantageous for some of the seed of a plant to forgo germination in the next year and risk germinating some years later when the environment may be more favourable. Most published research envisages a randomly varying environment and some fixed fraction G of the seed bank germinating each year. Simple life history models have been formulated which entail the existence of a value for G strictly less than unity ($G = 1$ is equivalent to lack of dormancy) which optimises some measure of long-term viability of the species.

All published work on this topic stems from a simple model advanced by Cohen (1966). Let S_n denote the number of seeds in the seed bank at the beginning of season n. Of this a proportion G germinates and grows to adult plants which yield Y_n seeds per plant. Here $\{Y_n\}$ is a sequence of independent random variables. Of the non-germinating component of the seed bank, a proportion V remains viable and survives till the next season (Cohen works with the proportion D which do not survive). Thus

$$S_{n+1} = S_n[GY_n + (1 - G)V].$$ (3.22)

Cohen's measure of long term viability is the long-term geometric mean fitness defined by $\lambda = \exp[\lim_{n \to \infty} (\log S_n)/n]$, provided this exists. He assumes also that the Y_n are identically distributed and then existence of λ is assured by the strong law of large numbers. Moreover

$$\lambda = \exp[E(\log(GY_1 + (1 - G)V))],$$

where $E(.)$ denotes the expectation operator. Cohen concentrates on the case where the Y_n can take only the values 0 and Y, a positive constant, corresponding to bad and good years, respectively. It is easily shown that, as a function of G, λ is maximised by

$$G_{max} = p - (1 - p)(1 - D)/(Y + D - 1)$$

where p is the probability that Y_1 assumes the value Y. The variation of λ and G_{max} as functions of D and Y is explored. Cohen also considers the case where Y_1 has a more complicated distribution, but here it is not possible to derive an explicit expression for G_{max} and Cohen explores necessary conditions for its existence.

In subsequent papers, Cohen considers variations of his simple model. Thus Cohen (1967) models the situation where the seed bank can receive partial information about the state of the environment and chooses its germination fraction accordingly. Specifically, year-to-year yields are random, as above, but now yearly viabilities are also random. In addition there are environmental indicators R_n such that if the current environmental state is $R_n = k$ then a fraction G_k of the seed bank will germinate. Thus (3.22) is replaced by

$$S_{n+1} = S_n[G_k Y_n + (1 - G_k)V_n].$$

Assuming that the random vectors (Y_n, V_n, R_n) are independent and identically distributed it follows that the geometric mean fitness exists. Cohen explores the existence and properties of a set of germination fractions which maximise the mean geometric fitness.

Cohen (1968) explores the modification of his basic model in which the yearly viabilities of ungerminated seeds are random. MacArthur (1972) gives an illuminating discussion of a slightly simplified version of Cohen's models ($V = 1$) in which he elucidates the model assumptions underlying the selection of G to maximise the mean geometric fitness or, alternatively, mean arithmetic fitness, that is, maximising $\lim_{n \to \infty} E[(S_n)^{1/n}]$. Also, see Leon (1985) for further discussion and ramifications of Cohen's models.

Klemow and Raynal (1983) report a field study which examines the extent to which the species *Erucastrum gallicum* is described by Cohen's models as opposed to what would be expected if it demonstrated phenotypic plasticity. A contrasting study of *Polygonum* had previously been conducted by Hickman (1975) who suggested that annual plants with dormant seeds growing in uncertain environments would exhibit, for example, a positive correlation between mean seed production and soil moisture, and high seedling survival in drought years. In contrast, under Cohen's scenario, seedling mortality should be high in drought years, surviving mature plants should be very fecund, seed germination should only marginally reduce the seed bank, and there should be evidence of substantial year to year seed survival. In Klemow and Raynal's study environmental variability is manifested in the frequency and intensity of rainfall.

They found that *Erucastrum g.* exhibited high plasticity by virtue of a highly variable seed production which ranged from 24 to 1675 seeds per plant. In most other respects this species behaved differently from that studied by Hickman and exhibited many of the traits predicted by Cohen. For example there was high seedling survival in moist years, complete plant mortality in drought years, high seed output compared to species in more stable sites, maintenance of the seed bank and seed dormancy lasting over one year.

Ritland (1983) investigated a modification of Cohen's basic model which allows for some plants to delay seed setting until later in the season. As before, S_n is the size of the seed bank just before germination in year n. A proportion G of these seeds germinate and grow into mature plants. A proportion $1 - a$ of mature plants set seed early in the season, yielding Y_n seed per plant. The remaining fraction of mature plants set their seed later in the season, yielding $Z_n Y_n$ seeds per plant where Z_n is a random factor. A random proportion V_n of seed left in the seedbank survives until the following year. Thus (3.22) is now replaced by

$$S_{n+1} = S_n[(1 - a)GY_n + aGZ_nY_n + (1 - G)V_n].$$

$$= S_n V_n[1 + GR_{1,n} + aGR_{2,n}]$$

where $R_{1,n} = (Y_n - V_n)/V_n$ and $R_{2,n} = (V_n - 1)Y_n/V_n$.

Assuming that the $R_{i,n}$ have positive expectation and variance, Ritland seeks to maximise the long-term geometric mean fitness $E[\log(V_1(1 + GR_{1,1} + aGR_{2,1})]$. This quantity is not directly calculable and hence Ritland uses a Taylor expansion to find a more tractable approximation. He investigates the maximisation of this approximation with respect to G and a separately and jointly. The adequancy of his approximation is checked using simulations. Ritland extends his analysis and discussions to allow for environmental prediction by plants and he also introduces a genetic structure.

The models above ignore the possibility of density dependent seedling loss. Models which take this into account typically predict bounded population sizes as a function of time, and even some kind of long-term equilibrium behavior. Under such a regime it is not appropriate to optimise long-term growth rates – they are zero. To surmount this some investigators have sought optimality conditions in terms of evolutionarily stable strategies (Maynard Smith (1982) and Hines (1987)).

This idea was applied by Bulmer (1984) to the evolution of seed dormancy. Very roughly, an established species has a germination fraction which is optimal in this sense if another plant species having identical characteristics (that is, dormant seed viability and density dependent seedling viability) but a different germination fraction cannot successfully invade the established species. Using primes to denote the invading species, (3.22) is replaced by the system

$$S_{n+1} = S_n[GYI_nf(GS_n + G'S'_n) + (1 - G)V]$$

and

$$S'_{n+1} = S'_n[G'YI_nf(GS_n + G'S'_n) + (1 - G')V'].$$

Here S_n and S'_n represent the seed numbers in year n of the established and invading species as they grow together in the same habitat; $I_n = 1$ if year n is a good one, and $I_n = 0$ otherwise; Y is the per plant seed yield of each species in a good year and $f(GS_n + G'S'_n)$ is the proportion of seedlings of each species which survive to maturity. When the invader is rare this function can be replaced by $f(GS_n)$ in the system above, giving a first order equation for S_n. Bulmer states that the second equation above yields a mean geometric fitness for the invading species given by

$$\lambda = E[\log\{G'YI_nf(GS_n) + (1 - G')V'\}]$$

where, presumably, the expectation is with respect to a putative stationary distribution of $\{S_n\}$. The condition that G be an evolutionarily stable optimum is that λ have a local maximum at $G' = G$. This condition leads to a rather intractable equation for the optimal value, but some numerical results can be determined for special forms of the function f. Bulmer ends his discussion with some remarks on spatial dispersion of seeds, the next topic on our agenda.

Just as seed dormancy, or temporal dispersal of seed, can be advantageous for annual plants, so also can spatial dispersal of seed be advantageous. We suppose that the habitat of a certain species is composed of many patches and environmental conditions vary from patch to patch as well as temporally. Thus a mature plant living in a patch currently experiencing deleterious conditions would gain an advantage if some of its offspring seed can be dispersed to patches with a more benign environment.

Venable and Lawlor (1980) consider a model for dispersal based on the simplest model in Cohen (1966) and MacArthur's (1972) discussion of it. Thus years are classified as good or bad and, as for Cohen's model, within each patch temporal environmental fluctuations are governed by a sequence of Bernoulli trials, and in addition these sequences are independent and identical in distribution across patches. Their model specifies a proportion D of the yield of any one plant which is dispersed to other patches, and this dispersal occurs randomly. The model formulation is in terms of the proportion D' of the total seed load of a patch which disperses. It can be shown that $D' = pGYD/[1 + G(pY - 1)]$.

As above, the object is to maximise the long-term mean fitness with respect to G. This fitness has two components. The first of these is just Cohen's geometric mean fitness

$$\lambda_c = (GY + 1 - G)^p (1 - G)^{1-p}$$

where, as above, Y is the plant yield in a good year and p is the probability that the current year is a good one. This component arises from the direct contribution to next year's seed bank in a target patch from its seed bank this year. This component must be discounted by the factor $1 - D'$ to allow for seed lost by dispersal.

The second component is MacArthur's arithmetic mean fitness

$$\lambda_m = p(GY + 1 - G) + (1 - p)(1 - G).$$

This is the expected yield from a single seed over one year and it is relevant here because the total contribution from dispersal to the target patch arises from the contributions of a large number of independently acting patches. Accordingly the law of large numbers says that the average per seed contribution of all other patches to the target patch is close to λ_m. This component must be discounted by D', leading to the total mean fitness $\lambda = (1 - D')\lambda_c + d'\lambda_m$. Venable and Lawlor maximise this quantity with respect to the germination fraction G. Denote the optimal value by G_{max}. They find that when p is small, G_{max} is small when D is small but rises very rapidly with modest increases in D. If D is moderate, $D \geqslant 0.3$ say, then G_{max} is fairly large ($\geqslant 0.75$)

irrespective of the value of p. Venable and Lawlor extend their discussion to take account of predictive germination.

In a long memoir, Levin, Cohen, and Hastings (1984) consider a more exact model which includes the effects of density dependent yield reduction. There are L patches and $S(i, n)$ is the number of seeds in patch i after dispersal into year n. A fraction G of these germinate with per capita yield $Y(i, n) = K(i, n)F(GS(i, n))$, where $K(i, n)$ represents the yield of a solitary plant in patch i in year n, and $F(.)$ represents the effects of density dependent yield reduction. The $K(i, n)$ are random variables which are uncorrelated in space and time. After seed are set in a given patch, a fraction D disperses and a fraction $1 - \alpha$ of these are lost. The remainder disperse independently and uniformly amongst all patches. Non-germinating seeds have mortality $1 - V$. Thus year-to-year seed numbers are described by the system

$$S(i, n + 1) = S(i, n)[GY(i, n)(1 - D) + (1 - G)V]$$

$$+ (\alpha DG/L)\sum S(j, n) Y(j, n). \qquad (3.23)$$

Assume now that there are several types of plant growing together with common values of α, G and V. The types are indexed by ℓ, so that $S^\ell(i, n)$ denotes the number of ℓ-type seeds in patch i at time n. The type specific parameters are denoted by D^ℓ and $Y^\ell(i, n)$ and it is assumed that density dependent yield reduction is described by

$$Y^\ell(i, n) = K(i, n)F(z_n)$$

where $z_n = G\sum_\ell S^\ell(i, n)$ is the total number of mature plants in patch i. Consequently (3.23) is replaced by a set of similar equations indexed by ℓ.

The object now is to determine an optimal choice of D, and owing to density dependence optimality is defined in terms of an evolutionarily stable strategy. The determination of an optimal value of D is simplified by two further assumptions. First, it is assumed that L is very large, thus reducing the problem to a deterministic one so far as patch selection variability is concerned. The second assumption is an ergodic hypothesis requiring long-term averages within patches to equal spatial averages evaluated at any time. Hardin, Taká, and Webb (1988) discuss conditions which entail this assumption. The evolutionarily stable strategy is obtained by requiring the patch seedbanks over all types to be in statistical equilibrium. This imposes a patch balance condition, namely, the nett inflow by dispersal of seeds into a patch must be exactly balanced by the number leaving by dispersal in the following year. At the individual level this requires the expected yield of each dispersed seed to be unity, and this condition leads to an equation for the optimal dispersal fraction. We will not summarize this paper any further, but refer the interested reader to it for a detailed discussion of the properties and consequences of the optimal strategy, and for various extensions of the basic model.

Klinkhamer et al. (1987) consider a density independent version of the model proposed by Levin et al. (1984). They argue that many annual plants grow in such harsh environments that they are never subject to density dependent yield reduction. Moreover this has the advantage of singling out the long-term geometric mean fitness as the natural optimality criterion. The general set up of Levin et al. is assumed to hold so that (3.23) continues to hold but now it is assumed that the yields $Y(i, n)$ are independent in time but may be correlated between patches. They develop an approximation to the geometric mean fitness in terms of the covariance structure of the $Y(i, n)$ and the other model parameters, but even this is not susceptible to a direct analysis. A combination of analytical and numerical methods are used to explore properties of the optimal dispersal and germination rates. The authors confirm and extend the insights gained in earlier investigations. For example it is found that the optimal dispersal fraction increases with increasing germination fractions. In the absence of dormancy or between patch correlation, optimal dispersal rates tend to be small, and this remains unaltered when these effects are allowed. Finally, the optimal germination fraction is very sensitive to changes in the dispersal fraction when this fraction is small. These findings imply that plant species with poor dispersing abilities have a higher dormancy than species with good dispersing abilities. Klinkhamer et al. present field data supporting this hypothesis.

Theoretical studies of the evolution of seed dormancy represents an active research area and papers continue to appear, extending previous investigations. For example, we mention a study by Venable and Brown (1988) of the evolutionary consequences of the three factors, spatial dispersal, dormancy, and seed size.

We turn now to a consideration of research into the dynamics of populations of plants having a seed bank, a topic which has received attention only recently. Generalising an earlier model of Watkinson (1980), MacDonald and Watkinson (1981) consider what they call a cross-over model which keeps track of the number N_n of mature plants in year n (that is, at the end of the growing season) and the size S_n of the seed bank just after all seed is set in year n. In all the models reviewed above it is assumed that seed germinates at a rate which is independent of its age. MacDonald and Watkinson take a first step in relaxing this assumption by distinguishing the germination rates of newly produced seed and the seed already in the seed bank, that is, seed aged more than one year.

An isolated mature plant produces a constant number λ of new seeds. The reduction of dry weight per plant due to competition is described through a function $F(N) \leqslant N$ such that the number of new seeds produced by N_n mature plants is $\lambda F(N_n)$. The new seed enters a dormant phase and a proportion Δ survive this. Of the surviving fraction, a proportion G germinates and a proportion $1 - G$ enters the seed bank. Finally, a proportion V of seedlings survive to maturity and hence $\lambda \Delta GVF(N_n)$ mature plants are produced in year $n + 1$ by the mature plants in year n.

At the end of the growing season in year n there are S_n seeds in the seed bank and a proportion Δ' of these survive the dormant phase. A fraction G' of these surviving seeds germinate and a proportion V of these survive to maturity. It follows that

$$N_{n+1} = \lambda \Delta GVF(N_n) + \Delta'G'VS_n.$$

The new seed which survives the dormant phase but does not germinate numbers $\lambda(1 - G)F(N_n)$ and a proportion V' of this survives the year $n + 1$ growth phase and is counted in the seed bank for this year. Similarly the contribution of the S_n seeds to the seed bank at the end of the year $n + 1$ growth phase is $\Delta'(1 - G')V'S_n$, and it follows that

$$S_{n+1} = \lambda \Delta V'(1 - G)F(N_n) + \Delta'V'(1 - G')S_n.$$

The factor λ is missing from the equations given by MacDonald and Watkinson (their (6) and (7)). Following Watkinson (1980), they use the function $F(N) = N/(1 + aN)^b$ where a and b are positive constants. Some properties of this pair of equations are elucidated using a local analysis and numerical examples.

MacDonald and Watkinson (1981) also consider a simplified version of the cross-over model which they call the bottleneck model. Their simplification is just that the new seed production is lumped with the existing seed bank. Mathematically this is tantamount to assuming $G = G'$ and $\Delta = \Delta'$. Letting $A = \Delta GV$ and $B = \Delta V'(1 - G)$ it is obvious from the equations above that $N_n = AS_n/B$ and hence the first equation simplifies to

$$N_{n+1} = \lambda AF(N_n) + BN_n.$$

MacDonald and Watkinson derive this equation by introducing an intermediate time step.

A stochastic version of the bottleneck model has been considered by Ellner (1984). He allows all the model parameters to vary randomly with time and he obtains conditions under which the mature plant numbers tend to zero or remain persistent with a limiting probability distribution. The model equation he uses is a little different to that above because he works with the size of the seed bank immediately prior to germination.

Ellner also considers a bottleneck version of competition between two annual plant species. He assumes that the seed production of each species is a function of the total number of seedlings and that the yield-density relationships is a binary version of Watkinson's function, used above, with $a = b = 1$. Thus the model equations used by Ellner are

$$X_{n+1} = \frac{\lambda_n G_n X_n}{1 + G_n X_n + G'_n Y_n} + \Delta_n (1 - G_n) X_n$$

and

$$Y_{n+1} = \frac{\lambda'_n G'_n Y_n}{1 + G_n X_n + G'_n Y_n} + \Delta'_n (1 - G'_n) Y_n$$

where λ_n, G_n and V_n are, for the first species and year n, the per plant fecundity, germination proportion and annual survivorship in the seed-bank, respectively, and the primed parameters denote the corresponding parameters for the second species. By simplifying his assumptions, Ellner is able to obtain a criterion which ensures the continued coexistence of the two species. Coexistence in this context means, for the first species say, that $\lim_{\varepsilon \to 0} \sup_n P\{X_n < \varepsilon\} = 0$.

A generalisation of the deterministic version of Ellner's equations has arisen in a model formulated by Jones and Perry (1978) for the numbers of two competing populations of cyst-nematodes under the assumption that the nematodes cause no damage to the root systems of their host plants. Their model gives rise to the pair of difference equations

$$p_{n+1} = (1 - \alpha) \frac{k_{12} p_n}{a + k_{12} p_n + q_n} + \alpha p_n$$

and

$$q_{n+1} = (1 - \beta) \frac{k_{21} q_n}{b + p_n + k_{21} q_n} + \beta q_n$$

where a and b are positive constants. This system cannot be put into the form of (3.7), although it might be expected that it will have similar properties to those we shall derive for (3.7). On the basis of intuition and simulation, Jones and Perry make several conjectures about the stability properties of their model and these are in accord with the properties of (3.7) derived in this monograph.

The possibility that germination rates of seed in the seed bank may depend on seed age is taken further by Templeton and Levin (1979) in their study of the time variation of the genetic structure of a population of annual plants. Their basic quantities are the proportions $\beta_n(i)$ of existing seeds which germinate in year n and were produced by adult plants i years earlier. These quantities are not the same as the probability $g_n(i)$ that in year n an i-year-old seed germinates. These quantities are related by

$$\beta_n(i) = g_n(i)S(n-i) \bigg/ \sum_{1 \leqslant j \leqslant N} g_n(j)S(n-j)$$

where N is the memory length and $S(n)$ is the total number of seeds produced in year n. The possibility that some seeds may never germinate is handled by allowing the possibility $\Sigma_j g_n(j) < 1$. Templeton and Levin seem to assume that the $S(n)$ can be specified independently of the germination probabilities, whereas they will be determined by these probabilities. However, they avoid the problem of determining the $S(n)$ by taking the $\beta_n(i)$ as given quantities. Finally, they assume unit seedling viability.

Organisms are assumed to have a single locus with alleles A and a, mutation is absent but selection may occur with fitnesses depending on absolute time. Difference equations are set up for $p(n)$, the frequency of the A allele among adult plants in year n, and $q(n) = 1 - p(n)$. In the case of constant fitnesses it is shown that there is an equilibrium A-allele frequency which depends only on the fitnesses. Thus seed dormancy has no effect on long-term allele frequencies.

Templeton and Levin claim, however, that seed dormancy does affect the rate at which genetic equilibrium is reached. They define a quantity Δp which is the amount of allele frequency change per year. Under the assumption of "weak selection" this quantity is assumed to be related to the $p(n)$ through $p(n) = p(n-i) + i\Delta p$. They display an approximation for Δp which depends on the age dependent germination rates. The derivation of this expression seems to contain a tacit assumption that the $p(n-i)$ which apparently occur in summations weighting the $\beta_n(i)$ [the notation is obscure at this point] are constant throughout the range of summation. If this is the case, then the relation defining Δp shows that it must be zero. Templeton and Levin go on to a consideration of the effects of varying environments.

Schmidt and Lawlor (1983) set up a model for a single species of annual plants having a number of features similar to those for our Model 2 which is analysed in Chapter 6. Thus new seedlings survive to adulthood with a constant probability and adult plants each produce Y

new seeds which enter the seedbank. The year is divided into winter and the summer growing season. Seed in the seed bank survives winter with a probability depending on its age and at the end of winter it germinates with a probability which depends on its age. Seed which does not germinate lies dormant and survives summer with a probability which is age dependent. Seed survival probabilities are taken to be zero for seed over two years old. The authors follow the number of new seedlings and the numbers of seeds which are aged 0, 1 or 2 at the start of summer and winter each year. The model specification allows winter–summer and summer–winter transitions to be described by winter and summer Leslie matrices, respectively. Year to year transitions are described by multiplying these matrices in the appropriate order. Since there are no seedlings at the beginning of winter it is easiest to work in terms of the sizes of the seed age classes corresponding to the onset of winter.

The authors give equations for the growth rate in this class and for the limiting age-class proportions. They give a lot of attention to a sensitivity analysis of the derived quantities and compare them with a model in which age classes are ignored. Inter alia, they find for both models that the growth rate is more sensitive to variations in adult plant fecundity and survivorship than to variations in germination fractions. In addition, the growth rate is more sensitive to changes in the seed summer survivorship parameters than in the corresponding winter parameters. These conclusions accord with those of Rossiter et al. (1985) and also with some of our findings in Chapter 7.

A comprehensive set of models has been devised by Pacala (1986b) which takes account of both dispersal and age dependent germination rates. It is probably easiest to begin by describing the family of models developed by Pacala and Silander (1985) for the growth of a single species of annual plants with no dormancy, and then describe how it is extended to account for multiplicity of species and dormancy.

The models are called neighborhood models because it is assumed that for each plant there is a circle, the neighborhood of this plant, which contains all other plants which interact directly with the given plant. Population growth of the species is assumed to be described by several predictors as follows.

The seedling fecundity predictor (SFP) predicts the future seed set of a seedling as a function of the numbers and locations of its neighboring seedlings. The adult fecundity predictor (AFP) gives the actual seed set of a mature plant as a function of the numbers and locations of its mature neighbours. The dispersal predictor (DP) predicts the dispersal pattern of a plant's seedling progeny and survivorship given the location and seed set of the mother plant. The DP incorporates information giving the survivorship from the seed set by the mother plant to its seedling progeny, and the disposition of the progeny seed around the mother. Finally, there is a survivorship predictor (SP) giving the probability that a given seedling survives to maturity given the number and location of its seedling neighbors.

Once analytical forms for these predictors are prescribed, they are used in two ways. First, it is possible to run a computer simulation which tracks seedling numbers and locations as a function of time (i.e., season number). Briefly, simple explicit forms of predictors which are common in the literature were used and it was assumed that the radii of all neighborhoods was the same; denote it by r. The simulation was run for twenty-five generations and it kept track of the locations of seedlings inside a square plot of side $13\frac{1}{3}r$ which was at the centre of a larger square plot of size $33\frac{1}{3}r$. Dispersal of individual seeds was assumed to be isotropic and the distance of dispersal from the mother plant was governed by an exponential distribution with mean $4r/3$.

Progeny seed of a given plant dispersed independently, and independently of the locations of the mother's neighbors. Seed hitting the boundary of the large square was assumed to be absorbed there. Finally, the seed yield of a given plant depended on only the number n of mature plants within the neighborhood of the parent and was a decreasing function of n. Two cases were considered: $Y(n) = Y/(1 + cn)$ and $Y(n) = Ye^{-cn}$ where Y is the yield of an isolated plant and c is a constant. This fully specifies the AFP used. Finally, a 100 percent germination rate was assumed and the SP was assumed to be a constant. This completely determines the simulation model.

The simulation was run for fixed values of all constants except that Y was varied over several runs. It was found for the hyperbolic yield function that seedling population densities always tended to a positive limiting value with little deviation from monotonicity, and the same was true for the exponential yield model provided Y was not too large. For larger Y it was found that the seedling densities fluctuated unpredictably in time. We will not write further about the simulation model because Pacala's (1986b) work on dormancy is entirely in terms of analytical models.

For the analytical models it is assumed that seedlings grow on an infinite plane and that progeny seed from a given plant disperses independently according to a spatial Poisson process. Thus if S_n is the average population density of seed in generation n and G is the rate of germination of seed, then the number of year $n + 1$ daughter seedlings in a region A has a Poisson distribution with mean $GS_n|A|$, where $|A|$ is the area of A. A first order difference equation specifying the S_n

results from the continuity equation,

yield = [seedling density][average survivorship][average fecundity].

Symbolically this can be expressed as

$$S_{n+1} = GS_n Z(S_n) F(S_n).$$

We obtain the functions $Z(.)$ and $F(.)$ as follows.

Assume that all neighborhoods have the same radius with units chosen so that $\pi r^2 = 1$. Then the fraction of year n seedlings having m neighbors is $[\exp - (GS_n)][GS_n]^m/m!$. If the SP specifies that adult survival depends on only the number of neighbors, then it can be described by a function $z(m)$, the probability that a given seedling with m neighbors survives to maturity. It is clear then that

$$Z(S_n) = [\exp - (GS_n)] \sum_{m \geqslant 0} z(m)[GS_n]^m/m!.$$

Similarly, if adult plant yield depends only on the number m of seedling neighbors and it is given by $f(m)$ then

$$F(S_n) = [[\exp - (GS_n)]/Z(S_n)] \sum_{m \geqslant 0} z(m)f(m)(GS_n)^m/m!,$$

because for each m the coefficient of $f(m)$ is the fraction of adult plants in year $n + 1$ that had m seedling neighbors.

Pacala and Silander (1985) show that a similar difference equation can be formulated using the AFP instead of the SFP, as above. A general analysis of the equation for S_n is very difficult, although it is possible to give a local stability analysis. Also, it transpires that if $z(.)$ is taken to be constant and $f(.)$ is given special forms then the summations above can be performed to give difference equations which have received extensive analysis in the literature. In particular, these examples show that it is possible for $\{S_n\}$ to have a GAS equilibrium, or to converge to stable oscillations, or to exhibit chaotic behavior.

Pacala (1986a) has extended these ideas to multi-species models, paying most attention to two species. In this context predictors must be specified for each species but they are assumed to be independent of neighbor locations. Following the formulation of Pacala (1986b), let $S_{i,n}$ be the density of i-type seedlings in year n. The first order difference equation above is here replaced by the system

$$S_{i,n+1} = S_{i,n} G_i Y_i(\mathbf{S}_n(i))$$

where G_i is the probability that an i-type seed produced in year n will survive and germinate in year $n + 1$. Each i-type seedling has an associated j-type neighborhood which is a circle of radius r_{ij} centered on the seedling and containing all the j-type seedlings interacting with it. The number of such interacting seedlings is $S_{j,n}(i) = S_{j,n} 4\pi r_{ij}^2$ and $\mathbf{S}_n(i)$ is the v-vector of these numbers, where v is the number of species. Finally, the yield predictor specifies the mean per capita seed yield of an i-type seedling and is given by the function $Y_i(\mathbf{S}_n(i))$. This in turn is given by

$$Y_i(\mathbf{S}_n(i)) = \sum_{\mathbf{m} \geqslant 0} D_i(\mathbf{m}|\mathbf{S}_n(i)) f_i(\mathbf{m})$$

where $\mathbf{m}$ is a v-vector of non-negative integers, $f_i(\mathbf{m})$ is the seed set predictor for an i-type seedling whose neighbor numbers are given by $\mathbf{m}$ and $D_i(\mathbf{m}|\mathbf{S}_n(i))$ is the probability that an i-type seedling has neighbor numbers $\mathbf{m}$ arising from dispersal when the seedling numbers per neighborhood are described by $\mathbf{S}_n(i)$. When dispersal is described by a Poisson process of unit intensity,

$$D_i(\mathbf{m}|\mathbf{S}_n(i)) = \prod_{1 \leqslant j \leqslant v} \exp(-S_{j,n}(i))(S_{j,n}(i))^{m_j}/m_j!.$$

To add dormancy to this model a more complicated germination predictor is required. Let p_i be the probability that an i-type seed produced in season n survives to the start of season $n+1$. Let $p_{i,k}$ be the probability that an i-type seed which is k years old at the start of a season survives to the next season. Finally, let $g_{i,k}$ be the probability that a viable i-type seed which is k years old at the start of a season will germinate in the same season.

Let $s_{i,k,n}$ be the population density of k year old i-type seeds counted just before germination in year n. With Poisson dispersal it can be shown that

$$s_{i,0,n+1} = p_i S_{i,n} Y_i(\mathbf{S}_n(i)),$$

$$s_{i,k+1,n+1} = s_{i,k,n} p_{i,k}(1 - g_{i,k})$$

and the seedling density is given by

$$S_{i,n} = \sum_{k \geqslant 0} g_{i,k} s_{i,k,n}.$$

These equations define a dynamical system which in general will be difficult to analyse. In addition it has the practical disadvantage of being defined in terms of germination parameters which are hard to estimate. Pacala shows that more easily estimated parameters are given by $\gamma_{i,0} = p_i g_{i,0}$ and for $k > 0$, $\gamma_{i,k} = p_i g_{i,k} \prod_{0 \leqslant a < k} p_{i,a}(1 - g_{i,a})$. These are more accessible because $\gamma_{i,k}$ is the probability that an i-type seed germinates in its kth year. With these parameters the following equations for seedling density can be derived:

$$S_{i,n+1} = \sum_{k \geqslant 0} \gamma_{i,k} S_{i,n-k} Y_i(\mathbf{S}_{n-k}(i)).$$

This system does not define a dynamical system as understood in the previous chapter. We will see in Chapter 6 that the same situation occurs for our formulation of Model 2. In the finite memory case we can produce a dynamical system in a higher dimensional state space which contains our Model 2, and a similar strategy is possible for Pacala's model.

Let S_i denote equilibrium values of the seedling densities. Then the above system of equations yields

$$G_i Y_i(\mathbf{S}(i)) = 1$$

where $G_i = \Sigma_k \gamma_{i,k}$ is the probability that an i-type seed eventually germinates. The interesting feature of this system of equations is that for given values of the G_i the system does not depend on the values of the age dependent germination and survival probabilities. Put another way, for given G_i and $Y_i(.)$, equilibrium seedling densities are independent of the way in which germination success is partitioned within age classes. A consideration of an example shows that in general this is not true of the equilibrium seed numbers. Indeed, by varying the germination predictor, but keeping the G_i fixed, it is possible for the seed numbers to assume arbitrarily large equilibrium values.

Pacala (1986b) goes on to consider the single-species case, showing that dormancy-induced time lags affect the local stability of equilibria. He also considers a two-species situation in which a second species is invading an established species. He shows that the ability to invade successfully, or otherwise, depends on the age dependent parameters. Finally, he looks at these questions for specific empirically calibrated models for velvet leaf and pigweed communities.

We end this review with the following observations. The pair of differential equations

$$dp/dt = (1 - \alpha)[k_{12}p/(k_{12}p + q) - p]$$

and

$$dq/dt = (1 - \beta)[\rho k_{21}q/(p + k_{21}q) - q],$$

can be regarded as continuous analogues of the discrete system (3.7). When $k_{12}k_{21} = 1$, Schoener (1976) has proposed a slightly more general pair of equations for describing the interaction of two species subject to a form of energy-limited competition, which he calls exploitative. By a consideration of the directions of the flow of this system within regions bounded by its isoclines and the boundary of the state space, Schoener obtains stability results which are analogous to some of our results for the system (3.7) when $k_{12}k_{21} = 1$.

Gates and Westcott (1988) have explored the behavior of the related system obtained by omitting the subtracted terms (p and q) from the righthand sides of the above differential equations. In their model p and q represent the quantities of plant present per unit area within a growing season of a pair of competing species. For a pair of annuals with no seed banks, they use the solution of their differential equations to bridge the gulf between seasons. They do this by formulating a recurrence relation linking the proportions of ground area shared by the first species at the start of successive seasons. They show that it is possible to choose parameter combinations such that either one species eventually displaces the other irrespective of initial densities, or a similar outcome can occur but with the winner depending on the initial densities of the two species, or the two species coexist indefinitely with the proportions of shared ground area tending to positive limits. These outcomes are similar to those we will find for Model 1A.

4

Analysis of Model G – no interior equilibria

4.1   Introduction

As we mentioned in Section 3.3, Model G has no internal equilibria if one isocline lies completely above the other in $\mathscr{S}^0$, although they may intersect on $\partial\mathscr{S}$. Allowing such intersections means that the necessary conditions for absence of internal equilibria given in the last chapter, Equation (3.19), can be weakened to the form

$$1 \leqslant p_G \quad \text{and} \quad q_F \leqslant \rho \tag{4.1a}$$

or

$$p_G \leqslant 1 \quad \text{and} \quad \rho \leqslant q_F, \tag{4.1b}$$

where we recall that $p_G = \rho G'(0+)$ and $q_F = F'(0+)$. If equality holds in any of these conditions then at least one of the boundary equilibria is neutrally stable and we may get no guidance from Theorem 3.3. For example if $1 < p_G$ and $q_F = \rho$ then, in the absence of internal equilibria, $\mathscr{I}_G$ lies above $\mathscr{I}_F$ and Theorem 3.3 tells us that $(1, 0)$ is unstable and the proof of this theorem shows that $(0, \rho)$ is neutrally stable (see (3.15) and (3.17) for the definitions of $\mathscr{I}_F$ and $\mathscr{I}_G$). Thus, although we may conjecture that $(0, \rho)$ is AS, Theorem 3.3 cannot aid us in deciding this. We proceed as follows.

Let us fix attention on the case where $\rho > 0$ and $\mathscr{I}_G$ lies above $\mathscr{I}_F$ in $\mathscr{S}^0$, in which case (4.1a) is satisfied. A representative configuration is shown in Figure 4.1. As indicated in Figure 4.1 the isoclines partition $\mathscr{S}$ into three regions $\mathscr{R}_1$, $\mathscr{R}_2$ and $\mathscr{R}_3$. The region between the isoclines, $\mathscr{R}_2$, is understood to be closed. If $(p, q) \in \mathscr{S}^0$, denote by (p', q') the image of (p, q) under the dynamical system defined by eq. (3.6), that is

$$p' = (1 - \alpha)F\left(\frac{p}{p + q}\right) + \alpha p, \tag{4.2a}$$

$$q' = \rho(1 - \beta)G\left(\frac{q}{p + q}\right) + \beta q. \tag{4.2b}$$

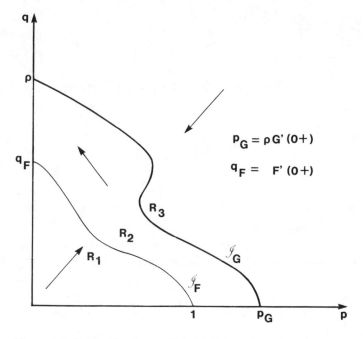

Figure 4.1. Typical isoclines of Model G when $\mathscr{I}_G$ lies above $\mathscr{I}_F$.

Now

$$(p' - p)/(1 - \alpha) = F(p/(p + q)) - p$$

and hence $p' - p$ has the same sign as the function

$$H(p, q) = F(p/(p + q)) - p.$$

Also, Equation (3.15) shows that (p, q) lies on $\mathscr{I}_F$ iff $H(p, q) = 0$. As $H(p, q)$ is strictly decreasing in q for each fixed p, it follows that $H(p, q) < 0$ iff $(p, q) \in \mathscr{R}_1$, and $H(p, q) > 0$ iff $(p, q) \in \mathscr{R}_2 \cup \mathscr{R}_3$. In particular we obtain the following trichotomy:

$$p' > p \text{ iff } (p, q) \in \mathscr{R}_1$$

$$p' = p \text{ iff } (p, q) \in \mathscr{I}_F$$

$$p' < p \text{ iff } (p, q) \in \mathscr{R}_2 \cup \mathscr{R}_3 \backslash \mathscr{I}_F.$$

A similar discussion applies to the sign of $q' - q$ in respect of the location of (p, q) relative to $\mathscr{I}_G$. In particular

$$p' > p \text{ and } q' > q \text{ if } (p, q) \in \mathscr{R}_1,$$

$$p' < p \text{ and } q' < q \text{ if } (p, q) \in \mathscr{R}_3,$$

and

$$p' < p \text{ but } q' > q \text{ if } (p, q) \in \mathscr{R}_2.$$

Thus, if a point of an orbit lies below both isoclines then the next point carries the orbit away from the origin in a north-easterly direction. Similarly, segments of an orbit lying in $\mathcal{R}_3$ drift in a south-westerly direction, and finally, segments inside $\mathcal{R}_2$ drift in a north-westerly direction, apparently toward the equilibrium $(0, \rho)$. These various directions of drift are indicated by the arrows in Figure 4.1.

These general characteristics of the orbits depend only on our assumption that $\mathcal{I}_G$ lies above $\mathcal{I}_F$ in $\mathcal{S}^0$ and hence they persist even if equality obtains in both the inequalities (4.2a). It appears, therefore, that any orbit starting in $\mathcal{S}^0$ will converge to $(0, \rho)$. We shall show in the next section that this *is* the case. The proof of this result allows us to obtain greater insight into orbit behavior. Consider a point $(p, q) \in \mathcal{S}^0$ on some orbit and its image (p', q'). If $r' = q'/p' > r = q/p$ then the orbit has jumped from a certain ray through the origin to another of larger gradient. Let $\mathcal{R}_q = \{(p, q) \in \mathcal{S}^0 : r' > r\}$ and $\mathcal{R}_p = \{(p, q) \in \mathcal{S}^0 : r' < r\}$. This notation is used to suggest that if $(p, q) \in \mathcal{R}_q$ then the orbit through (p, q) is locally spiralling toward the q-axis. The proof we shall give of Theorem 4.1 suggests that when they are non-empty the sets $\mathcal{R}_p$ and $\mathcal{R}_q$ have simple structures and in particular they are connected. The remainder of Section 2 will be devoted to determining the form of these sets as far as is possible for Model G. In Section 3 we concentrate on Model 1A where we can obtain a complete picture of the shapes of $\mathcal{R}_p$ and $\mathcal{R}_q$. This section contains many details which are not required for the sequel.

Consider an orbit with initial point in $\mathcal{R}_3^0$. Given that this orbit converges to the equilibrium point $(0, \rho)$, it is not clear whether the final approach to this point is from within $\mathcal{R}_2$ or not. In other words, is it the case that every orbit starting within $\mathcal{R}_3$ must eventually cross the isocline $\mathcal{I}_G$? One way of tackling this question is to determine the exact rates at which $\{p_n\}$ and $\{\rho - q_n\}$ approach zero. This will be the topic of Section 4 of this chapter. We shall see in the case of Model 1A that orbits may approach $(0, \rho)$ without entering $\mathcal{R}_2$, although this does not happen for the parameter values estimated from the experiments reported in Chapter 7. Finally, in Section 5 we will examine the degenerate case of Model G wherein $\rho = 0$.

4.2 Global asymptotic stability

As mentioned in Section 1 we will think in terms of the situation illustrated in Figure 4.1 where, in $\mathcal{S}^0$, the q-isocline, $\mathcal{I}_G$, lies above the p-isocline $\mathcal{I}_F$. Our object in this section is to prove the following result.

Theorem 4.1.

Suppose $\rho > 0$ and $\mathcal{I}_G$ lies above $\mathcal{I}_F$ within $\mathcal{S}^0$. Then $\mathcal{R}_2$, $\mathcal{R}_1 \cup \mathcal{R}_2$ and $\mathcal{R}_2 \cup \mathcal{R}_3$ are positively invariant. In addition the point $(0, \rho)$ is GAS. If $\mathcal{I}_F$ lies above $\mathcal{I}_G$ in $\mathcal{S}^0$ then the above positive invariance holds and then $(1, 0)$ is GAS.

Given the orbital drift directions shown in Figure 4.1 the reader may wonder why any further proof is required that $(0, \rho)$ is GAS. The reason is that because the orbits evolve discontinuously, we must entertain the possibility that they may jump out of $\mathcal{R}_2$ into $\mathcal{R}_3$, or jump directly from $\mathcal{R}_1$ to $\mathcal{R}_3$, or vice versa. Such erratic behavior would not be inconsistent with the drift directions indicated in Figure 4.1. The core of the proof of Theorem 4.1 is in showing the impossibility of such behavior, that is, in proving the positive invariance assertions.

Recall that if $(p, q) \in \mathcal{S}^0$ then we define $z = p/(p + q)$. It follows that the ray from the origin through the point (p, q) has gradient $q/p = (1 - z)/z$. Denote this ray by R_z and observe that for each $z \in (0, 1)$ it meets both isoclines (exactly once – see the remarks following Lemma 5.1). Let $\mathbf{x} = (p, q) \in R_z \cap \mathcal{I}_F$ and $\tilde{\mathbf{x}} = (\tilde{p}, \tilde{q}) \in R_z \cap \mathcal{I}_G$. We say that $\mathcal{I}_G$ lies above $\mathcal{I}_F$ in $\mathcal{S}^0$ if $\|\mathbf{x}\| < \|\tilde{\mathbf{x}}\|$ for each $z \in (0, 1)$. We then have $F(z) = p < \tilde{p} = \tilde{q}z(1 - z)^{-1}$, that is,

$$F(z) < \frac{z}{1 - z}\rho G(1 - z) \quad (0 < z < 1). \tag{4.3}$$

Conversely, (4.3) implies that $p < \tilde{p}$ if $\mathbf{x} \in R_z \cap \mathcal{I}_F$ and $\tilde{\mathbf{x}} \in R_z \cap \mathcal{I}_G$, and hence $q < \tilde{q}$. This gives the following criterion.

Lemma 4.1.

Inequality (4.3) is necessary and sufficient for $\mathcal{I}_G$ to lie above $\mathcal{I}_F$ in $\mathcal{S}^0$. The reverse of (4.3) is necessary and sufficient for $\mathcal{I}_F$ to lie above $\mathcal{I}_G$ in $\mathcal{S}^0$.

Proof of Theorem 4.1.

Let $(p, q) \in \mathcal{S}^0$, z be as above and $z' = p'/(p' + q')$. Algebraic manipulation of (4.1) reveals that

$$\text{sgn}(z' - z) = \text{sgn}\left[(1 - \alpha)F(z) + (\alpha - \beta)p - (1 - \beta)\frac{\rho z}{1 - z}G(1 - z)\right].$$

$$\tag{4.4}$$

Let $\sigma(p, q)$ be the term in square brackets.

Suppose now that $(p, q) \in \mathcal{R}_2 \cap \mathcal{S}^0$, in which case $p \geqslant F(z)$ and hence, from (4.3),

$$\sigma(p, q) \leqslant (1 - \alpha)p + (\alpha - \beta)p - (1 - \beta)(p/q)\rho G(1 - z)$$
$$= (1 - \beta)(p/q)(q - \rho G(1 - z)) \leqslant 0.$$

However, at least one of the inequalities giving the final bound is strict whence

$$z' < z \quad \text{if} \quad (p, q) \in \mathcal{R}_2 \cap \mathcal{S}^0. \tag{4.5}$$

It follows then, from (4.5) and the monotonicity of $F(.)$, that

$$p' - F(z') = (1 - \alpha)F(z) + \alpha p - F(z')$$
$$= F(z) - F(z') + \alpha(p - F(z)) \geqslant 0.$$

In a similar manner, $(p, q) \in \mathcal{R}_2 \cap \mathcal{S}^0$ entails $q \leqslant \rho G(1 - z)$ and hence $q' \leqslant \rho G(1 - z')$. Thus $(p', q') \in \mathcal{R}_2 \cap \mathcal{S}^0$, that is, this set is positively invariant. That $\mathcal{R}_2$ is also positively invariant will follow from a consideration of orbits which start on $\partial \mathcal{S}$. For example if $p = 0$ then (4.2b) becomes $\rho - q' = \beta(\rho - q)$ and hence if $q_F \leqslant q \leqslant \rho$ then $q_F < q' \leqslant \rho$, that is, if $(0, q) \in \mathcal{R}_2$ then $(0, q') \in \mathcal{R}_2$.

The proof of invariance of $\mathcal{R}_1 \cup \mathcal{R}_2$ and of $\mathcal{R}_2 \cup \mathcal{R}_3$ also depends on the relation between z' and z, and this depends on the relation between α and β. We shall only discuss the case $\alpha \geqslant \beta$ and leave the converse case to the reader.

If $\alpha = \beta$ it follows directly from (4.3) and (4.4) that $z' < z$ everywhere in $\mathcal{S}^0$. In all cases we have

$$\sigma(p, q) < (1 - \alpha)F(z) + (\alpha - \beta)p - (1 - \beta)F(z)$$
$$= (\alpha - \beta)(p - F(z))$$

and the last term is negative if $\alpha > \beta$ and $(p, q) \in \mathcal{R}_1$ because in that case $p < F(z)$. Hence we still have $z' < z$ and it follows that $q < \rho G(1 - z)$ entails $q' < \rho G(1 - z')$. We conclude that $\mathcal{R}_1 \cup \mathcal{R}_2$ is positively invariant.

Now suppose that $(p, q) \in \mathcal{R}_3 \cap \mathcal{S}^0$. We shall shortly see that it is possible for $z' \geqslant z$ when $\alpha > \beta$. Suppose first, however, that $z' < z$. The reasoning used for $\mathcal{R}_2$ shows that $p > F(z)$ entails $p' > F(z')$. On the other hand, if $z' \geqslant z$ then

$$q' - \rho G(1 - z') = (1 - \beta)(G(1 - z) - G(1 - z'))$$
$$+ \beta(q - \rho G(1 - z)).$$

Obviously the first term is non-negative and the second term is positive, since $(p, q) \in \mathcal{R}_3^0$. It follows that $(p', q') \in \mathcal{R}_3^0$ and again we conclude that $\mathcal{R}_2 \cup \mathcal{R}_3$ is positively invariant.

We now conclude the proof by showing that $(0, \rho)$ is GAS. Suppose $\mathbf{x} \in \mathcal{R}_1^0$. The orbit $\{(p_n, q_n)\}$ starting from $\mathbf{x}$ can never enter $\mathcal{R}_3$ and hence $\{q_n\}$ is a strictly increasing and bounded sequence, whence $Q = \lim_{n \to \infty} q_n$ exists and $0 < Q < \infty$. Initially, the sequence $\{p_n\}$ is also strictly increasing. Thus the orbit $\{\mathbf{x}_n\}$ starting from $\mathbf{x}$ jumps toward $\mathscr{I}_F$ and it jumps into $\mathscr{R}_2 \cap \mathscr{S}^0$ in finitely many steps. To see this, observe that in the converse case $\{p_n\}$ would also be a strictly increasing bounded sequence and hence $\{\mathbf{x}_n\}$ would converge to a point $\mathbf{a} \in \mathscr{S}^0$ and this limit point must be an interior equilibrium. But we suppose there are no such points.

Thus there exists ν such that $\mathbf{x}_\nu \in \mathscr{R}_2 \cap \mathscr{S}^0$, and since $\mathscr{R}_2$ is positively invariant it follows that $\{p_n : n \geqslant \nu\}$ is strictly decreasing – the orbit jumps towards the q-axis. It follows that the orbit converges to an equilibrium point in $\mathscr{R}_2$ and the only possible candidate is $(0, \rho)$.

If $\mathbf{x} \in \mathscr{R}_2 \cap \mathscr{S}^0$ then the orbit starting at $\mathbf{x}$ converges to $(0, \rho)$ and if $\mathbf{x} \in \mathscr{R}_3^0$ then the orbit enters $\mathscr{R}_2 \cap \mathscr{S}^0$ or it stays in $\mathscr{R}_3^0$. In either case it must converge to $(0, \rho)$. This completes the proof.

We shall now try to determine the regions $\mathscr{R}_p$ and $\mathscr{R}_q$ defined in the last section. Observe that the locus of constant z is the ray $q = z^{-1}(1 - z)p$ whose gradient increases as z decreases within $(0, 1)$. It follows that $\mathscr{R}_q = \{(p, q) \in \mathscr{S}^0 : z' < z\}$ and hence we can discover something about $\mathscr{R}_p$ and $\mathscr{R}_q$ by pursuing some of the ideas in the proof of Theorem 4.1. As we mentioned in Section 1, if $\mathbf{x} \in \mathscr{R}_q$ then a segment of the orbit through $\mathbf{x}$ is spiralling toward the q-axis. Alternatively we may think of that segment as being attracted by the GAS equilibrium point $(0, \rho)$. This is always the case if $\mathbf{x} \in \mathscr{R}_2 \cap \mathscr{S}^0$ and also if $\mathbf{x} \in \mathscr{R}_1^0$ when $\alpha \geqslant \beta$ or if $\mathbf{x} \in \mathscr{R}_3^0$ and $\alpha \leqslant \beta$. If $\alpha > \beta$ then for some $\mathbf{x} \in \mathscr{R}_3^0$ we can have $z' > z$, that is, $\mathbf{x} \in \mathscr{R}_p$ and we may then think in terms of an orbital segment through $\mathbf{x}$ being attracted by the unstable equilibrium point $(1, 0)$. Eventually, of course, every orbit enters $\mathscr{R}_q$.

Suppose that the hypotheses of Theorem 4.1 are in force and that $\alpha > \beta$. Let $\mathscr{C} = \{(p, q) \in \mathscr{S}^0 : z' = z\}$. In words, if an orbital point falls on $\mathscr{C}$, the orbital segment defined by $\mathbf{x}$ and $\mathbf{x}'$ lies in R_z, that is, this portion of the orbit spirals toward neither of the equilibria. As we shall see, as an orbit crosses $\mathscr{C}$ it comes under the influence of the GAS point. If both $\mathscr{R}_p$ and $\mathscr{R}_q$ are non-empty then $\mathscr{C}$ is the intersection of their closures and the discussion in the proof of Theorem 4.1 shows that $\mathscr{C} \subset \mathscr{R}_3$. It follows from (4.4) that $\mathbf{x} \in \mathscr{C}$ iff $\sigma(p, q) = 0$ which can be expressed as

$$(1 - \alpha)F(z)/z + (\alpha - \beta)(p + q) = (1 - \beta)\rho G(1 - z)/(1 - z).$$

$$(4.6)$$

The terms involving z are bounded on $[0, 1]$ and hence as $\mathbf{x}$ traverses $\mathscr{C}$, $p + q$ remains bounded. It follows that $\mathscr{C}$ is bounded.

For any $z \in (0, 1)$ it follows from (4.3) and our assumption $\alpha > \beta$ that $(1 - \beta)\rho G(1 - z)/(1 - z) > (1 - \alpha)F(z)/z$ and hence the equation

$\sigma(p, q) = 0$ can be solved for p. Denoting the solution by $p(z)$, we see
that $p(z) > 0$ and the corresponding value of q is $q(z) = z^{-1}(1 - z)p(z)$.
Thus $\mathscr{C} = \{p(z), q(z)): 0 < z < 1\}$.

As $z \to 0$ we see that $p(z) \to 0$ whence, from (4.6), $q(z) \to q(0)$ where

$$q(0) = \frac{1 - \beta}{\alpha - \beta}\left(\rho - \frac{1 - \alpha}{1 - \beta}q_F\right) > \rho. \tag{4.7a}$$

Similarly, as $z \to 1$, we have $q(z) \to 0$ and $p(z) \to p(1)$ where

$$p(1) = \frac{1 - \beta}{\alpha - \beta}\left(p_G - \frac{1 - \alpha}{1 - \beta}\right) > p_G. \tag{4.7b}$$

We conclude that $\mathscr{C}$ is nonempty and is a smooth curve lying in $\mathscr{R}_3$ and
joining the points $(0, q(0))$ and $(p(1), 0)$. Also, $\mathscr{R}_q$ is the open set
bounded by $\mathscr{C}$, the segment $[0, q(0)]$ of the q-axis and the segment
$[0, p(1)]$ of the p-axis, see Figure 4.2. This figure also shows the typical
behavior of some orbits. An orbit starting in $\mathscr{R}_p$ drifts towards the
origin and also spirals toward the p-axis. When it crosses $\mathscr{C}$ it does so
nearly tangentially to the ray containing the last point before it jumps
into $\mathscr{R}_q$. It continues drifting toward the origin but spiraling now

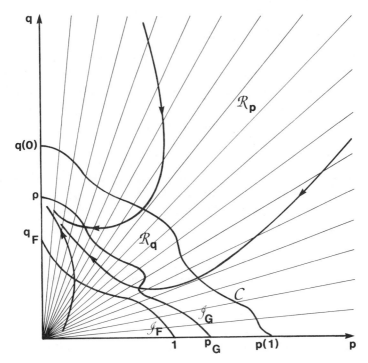

Figure 4.2. Typical configuration of some orbits of Model G when
$(0, \varrho)$ is GAS and $\alpha > \beta$.

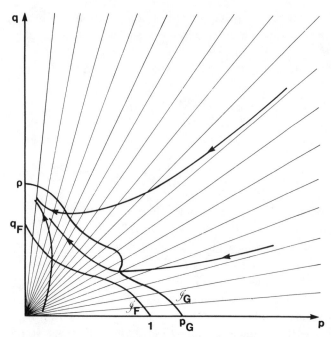

Figure 4.3. Typical configuration of some orbits of Model G when $(0, \rho)$ is GAS and $\alpha = \beta$.

toward the q-axis. Eventually it hops over $\mathscr{I}_G$, in a direction nearly parallel to the p-axis, since $q' \approx q$ if (p, q) is near $\mathscr{I}_G$, and then spirals up toward $(0, \rho)$. Orbits starting in $\mathscr{R}_1$ always spiral toward the q-axis and cross $\mathscr{I}_F$ into $\mathscr{R}_2$ in a direction almost parallel to the q-axis.

As a final observation note that as α tends to β from above both $p(1)$ and $q(0)$ tend to infinity. Indeed the curve $\mathscr{C}$ drifts away from the origin and $\mathscr{R}_q \to \mathscr{S}^0$. Figure 4.3 illustrates the situation when $\alpha = \beta$.

When $\alpha < \beta$ the situation is much less clear. The working in the proof of Theorem 4.1 shows that if $\mathbf{x} \in (\mathscr{R}_2 \cup \mathscr{R}_3)^0$ then $z' < z$. If $\mathbf{x} \in \mathscr{R}_1^0$ we cannot exclude the possibility that $z' < z$. Thus if $\mathscr{R}_p \neq \varnothing$ then $\mathscr{R}_p \subset \mathscr{R}_1^0$. Write (4.6) in the form

$$F(z) - \frac{\beta - \alpha}{1 - \alpha} p = \frac{1 - \beta}{1 - \alpha} \frac{\rho z}{1 - z} G(1 - z).$$

If β is sufficiently close to α it follows from (4.3) that this equation does not have a positive solution for p for any $z \in (0, 1)$ and consequently $\mathscr{R}_q$ coincides with $\mathscr{S}^0$. On the other hand, if β is close to unity there may be a solution with $p > 0$ for each $z \in (0, 1)$, or for z in some proper subset of $(0, 1)$. In order to resolve these problems it is necessary to specify F and G more precisely than heretofore. In the next section we shall consider the particular case of Model 1A.

4.3 Determination of $\mathscr{C}$ for Model 1A

We will simplify the notation a little by writing $c = k_{12}$ and $d = k_{21}$ so that

$$F(z) = \frac{cp}{cp + q} \quad \text{and} \quad G(1 - z) = \frac{dq}{dq + p}.$$

Equations (3.15) and (3.17), defining $\mathscr{I}_F$ and $\mathscr{I}_G$, respectively, reduce to the linear equations

$$\mathscr{I}_F : q = c(1 - p) \quad (0 \leqslant p \leqslant 1) \tag{4.8a}$$

and

$$\mathscr{I}_G : q = \rho - p/d \quad (0 \leqslant p \leqslant \rho d). \tag{4.8b}$$

We also have

$$p_G = \rho d, \quad q_F = c$$

and the criterion (4.3) for $\mathscr{I}_G$ to lie above $\mathscr{I}_F$ in $\mathscr{S}^0$ becomes

$$c \leqslant \rho \quad \text{and} \quad 1 \leqslant \rho d \tag{4.9}$$

and at least one inequality is strict. Conversely, $\mathscr{I}_F$ lies above $\mathscr{I}_G$ iff $c \geqslant \rho$ and $\rho d \leqslant 1$ and at least one inequality is strict. Three cases need consideration, namely, $cd = 1$, $cd > 1$ and $cd < 1$. We shall analyse the first two cases and leave the third for the reader.

(I). $cd = 1$. In this case the isoclines are parallel straight lines and $\mathscr{I}_G$ lies above $\mathscr{I}_F$ iff $c < \rho$. Then (4.6) yields

$$p\left[(1 - \alpha)\frac{c}{cp + q} - (\beta - \alpha) - (1 - \beta)\frac{\rho}{cp + q}\right] = 0$$

and if $p \neq 0$ it has the solution

$$q = q(0) - cp$$

where

$$q(0) = \frac{(1 - \alpha)c - (1 - \beta)\rho}{\beta - \alpha}.$$

If $\alpha > \beta$ we see directly that $q(0) > \rho$ and hence $\mathscr{C}$ is a straight line in $\mathscr{R}_3$ parallel to the isoclines. If $\alpha < \beta$ and

$$\frac{1 - \alpha}{1 - \beta} \leqslant \rho/c$$

then $q(0) \leqslant 0$ and hence $\mathscr{C}$ is empty. When this inequality is reversed then $\mathscr{C}$ is a straight line below and parallel to both isoclines, and $\mathscr{R}_p$ is that part of $\mathscr{S}^0$ below $\mathscr{C}$.

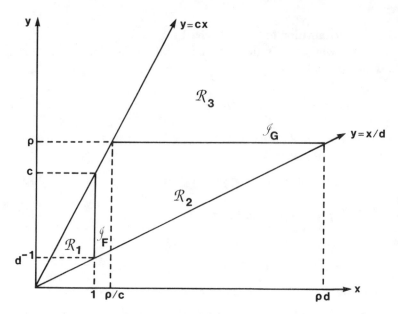

Figure 4.4. The transformed state space for Model 1A when $cd > 1$.

(II). $cd > 1$. When $cd > 1$ the first inequality at (4.9) yields $\rho > d^{-1}$, that is, strict inequality in the second of (4.9). Thus the isoclines can meet on the q-axis but not on the p-axis. The equation defining $\mathscr{C}$ is a general second degree equation which shows that $\mathscr{C}$ is a segment of some conic section. Greater insight can be gained as follows.

Define the linear transformation

$$x = (cp + q)/c \quad \text{and} \quad y = (p + dq)/d, \tag{4.10}$$

which has the inverse

$$p = d\frac{cx - y}{cd - 1} \quad \text{and} \quad q = c\frac{dy - x}{cd - 1}.$$

The transformation (4.10) maps $\mathscr{S}$ into an infinite wedge $\mathscr{W}$ bounded by the rays $y = cx$ and $y = x/d$, the first, the image of the q-axis, lying above the second. The equations for the isoclines, (4.8), transform into $x = 1$ and $y = \rho$. See Figure 4.4.

The equation for $\mathscr{C}$ is

$$(1 - \alpha)\frac{c}{cp + q} + (\alpha - \beta) - (1 - \beta)\frac{\rho d}{dq + p} = 0$$

which transforms to

$$\frac{1 - \alpha}{x} + \alpha - \beta - \frac{(1 - \beta)\rho}{y} = 0. \tag{4.11}$$

the equation of a rectangular hyperbola.

When $\alpha > \beta$ we see that

$$y \rightarrow \frac{(1-\beta)\rho}{\alpha - \beta} > \rho \quad \text{as} \quad x \rightarrow \infty.$$

This horizontal asymptote lies above $\mathscr{I}_G$ in $\mathscr{W}$ and maps into the straight line

$$q = \frac{(1-\beta)\rho}{\alpha - \beta} - p/d \tag{4.12}$$

in $\mathscr{S}^0$ which lies above, and parallel to, $\mathscr{I}_G$.

The hyperbola (4.11) has a vertical asymptote through the point $x = -(\alpha - \beta)/(1 - \alpha)$ and this maps back into the line

$$q = -c\left(\frac{\alpha - \beta}{1 - \beta} + p\right) \tag{4.13}$$

which lies below and parallel to $\mathscr{I}_F$. Solving (4.11) for y gives

$$y = \frac{(1-\beta)\rho x}{1 - \alpha + (\alpha - \beta)x}$$

and hence the limb of the hyperbola lying to the right of the vertical asymptote passes through the origin and lies below the horizontal asymptote. We conclude from this that $\mathscr{C}$ is a concave curve and is that part of the hyperbola with asymptotes (4.12) and (4.13), and it joins the points $(p(1), 0)$ and $(0, q(0))$ where

$$p(1) = \frac{(1-\beta)\rho d - (1 - \alpha)}{\alpha - \beta}$$

and

$$q(0) = \frac{(1-\beta)\rho - (1 - \alpha)c}{\alpha - \beta}; \tag{4.14}$$

see Figure 4.5. Observe that the asymptotes always intersect in the second quadrant when $cd > 1$. In addition, $\mathscr{C}$ lies between $\mathscr{I}_G$ and the segment in $\mathscr{S}$ of the asymptote defined by (4.12).

Suppose now that $\alpha < \beta$. The hyperbola (4.11) has a limb, denoted by $\mathscr{H}$, and defined by

$$y = \phi(x) = \frac{(1-\beta)\rho x}{1 - \alpha - (\beta - \alpha)x}$$

passing through the origin and lying to the left of its vertical asymptote at $x = (1 - \alpha)/(1 - \beta) > 1$. Observe that $\phi(1) = \rho$. The other limb lies to the right of this asymptote and *below* the horizontal asymptote

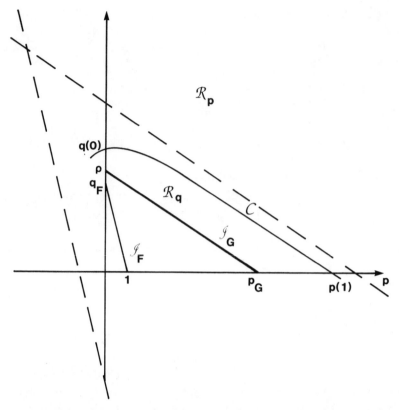

Figure 4.5. Model 1A with $cd > 1$ and $\alpha > \beta$. The broken lines are the asymptotes of the hyperbola defining $\mathscr{C}$.

through $y = -(1 - \beta)\rho/(\beta - \alpha) < 0$. The second limb never intersects $\mathscr{W}$ and is not relevant to our discussion.

Now

$$\phi'(0) = \frac{(1 - \beta)}{1 - \alpha}\rho \quad (<\rho)$$

and hence there are three cases to be considered:

(i) $c \leqslant \dfrac{(1 - \beta)}{1 - \alpha}\rho$;

(ii) $d^{-1} \leqslant \dfrac{1 - \beta}{1 - \alpha}\rho < c$; and

(iii) $\dfrac{1 - \beta}{1 - \alpha}\rho < d^{-1}$.

In case (i),

$$\phi(x) > cx \quad (x > 0),$$

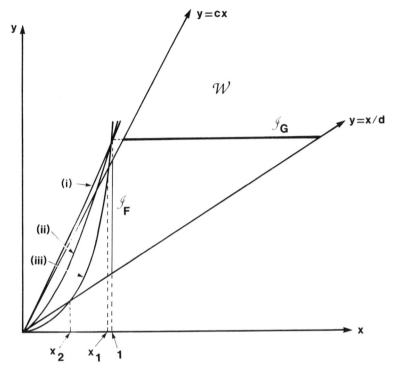

Figure 4.6. The transformed state space for Model 1A when $cd > 1$ and $\alpha < \beta$. The curves represent the hyperbola $\mathscr{H}$ in each of the cases (i)–(iii) discussed in the text.

that is, $\mathscr{H}$ does not intersect $\mathscr{W}$ and, in the $p - q$ plane, $\mathscr{C}$ is empty and $\mathscr{R}_q = \mathscr{S}^0$. In case (ii) (see Figure 4.6) $\mathscr{H}$ passes through $\mathscr{W}$ and leaves through the upper boundary at a point whose abscissa is

$$x_1 = q(0)/c = \frac{(1 - \alpha)c - (1 - \beta)\rho}{c(\beta - \alpha)}.$$

Clearly $\mathscr{C}$ is the image of the segment of $\mathscr{H}$ in $\mathscr{W}$ and is a convex curve joining the origin to $(0, q(0))$ and $\mathscr{R}_p$ is the open set whose boundary is $\mathscr{C}$ and the segment $[0, q(0)]$ of the q-axis. See Figure 4.7.

In case (iii) the tangent at the origin of $\mathscr{H}$ is less than the slope $1/d$ of the lower boundary of $\mathscr{W}$ whence $\mathscr{H}$ enters $\mathscr{W}$ at a point whose abscissa is

$$x_2 = \frac{1 - \alpha - (1 - \beta)\rho d}{\beta - \alpha}$$

and leaves $\mathscr{W}$ through a point whose abscissa is x_1, and $x_2 < x_1$. It is easy to check that

$$1 - x_1 = \frac{1 - \beta}{\beta - \alpha}(\rho/c - 1) \geqslant 0,$$

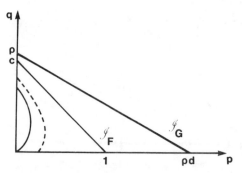

Figure 4.7. Model 1A with $cd > 1$ and $\alpha < \beta$. The curves represent $\mathscr{C}$ in Case (ii) (——) and Case (iii) (– – –).

and it should be clear that $\mathscr{C}$ is a convex curve joining the points $(0, q(0))$ and $(p(1), 0)$ and $p(1) = x_2$; see Figure 4.7.

Let us consider how $\mathscr{C}$ changes as ρ decreases from large values to c, its least possible value. If $\rho \geqslant \rho_1 = c(1 - \alpha)/(1 - \beta)$ we are in the domain of Case (i) and then $\mathscr{C}$ is empty. As ρ drops below ρ_1 we enter the domain of Case (ii) and the point $q(0)$ bifurcates from the origin and moves up the q-axis. Two sub-cases must be considered.

I. $cd \geqslant (1 - \alpha)/(1 - \beta)$. Since $\rho \geqslant c$, we also have $\rho d \geqslant (1 - \alpha)/(1 - \beta)$ and hence Case (iii) cannot occur. Referring to (4.14) we see that as ρ decreases from ρ_1 to c, $q(0)$ increases from 0 to c. It is not hard to show that the tangent to $\mathscr{C}$ through $(0, 0)$ has gradient

$$\sigma_0 = \frac{c}{d} \cdot \frac{\rho d(1 - \beta) - (1 - \alpha)}{c(1 - \alpha) - \rho(1 - \beta)}$$

and the tangent through $(0, q(0))$ has gradient

$$\sigma_1 = -\frac{c^2 d(1 - \alpha) - (1 - \beta)\rho}{d(c(1 - \alpha) - \rho(1 - \beta))}.$$

Observe that $\sigma_0 + \sigma_1 < 0$. As ρ decreases from ρ_1 to c, σ_0 decreases from ∞ to

$$\sigma_{0m} = \frac{c}{d} \cdot \frac{cd\rho(1 - \beta) - (1 - \alpha)}{c(1 - \alpha) - \rho(1 - \beta)}$$

which is 0 iff $cd\rho = (1 - \alpha)/(1 - \beta)$. At the same time σ_1 increases from $-\infty$ to

$$\sigma_{1M} = -\frac{c}{d} \cdot \frac{cd(1 - \alpha) - (1 - \beta)}{c(1 - \alpha) - \rho(1 - \beta)},$$

which satisfies $\sigma_{1M} < -c$ for any combination of parameters allowed by Case (I).

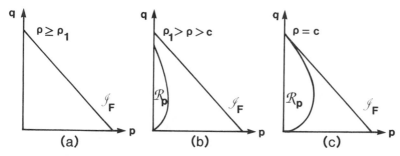

Figure 4.8. Model 1A: Growth of $\mathscr{C}$ as ρ decreases in Case I, $cd \geqslant (1 - \alpha)/(1 - \beta)$. (a) shows Case (i), where $\mathscr{C}$ is empty. (b) and (c) show Case (ii).

We conclude from this that as ρ decreases $\mathscr{C}$ grows rather like an expanding bubble adhering to the q-axis and sagging down toward the p-axis. When $\rho = c$ this bubble occupies the interval $[0, c]$ on the q-axis, that is, it touches $\mathscr{I}_F$ at its point of intersection with the q-axis. It can touch the origin tangentially to the p-axis, but when it touches $\mathscr{I}_F$ it can never touch it tangentially. See Figure 4.8.

II. $1 < cd < (1 - \alpha)/(1 - \beta)$. As ρ decreases toward c it passes first through ρ_1, then through $\rho_2 = d^{-1}(1 - \alpha)/(1 - \beta)$, and $\rho_1 > \rho_2 > c$. The behavior of $p(0)$ and σ_1 is similar to their behavior in Case I. As ρ decreases below ρ_1, $p(0)$ bifurcates from the origin and increases to $(c - d^{-1})(1 - \alpha)/(\beta - \alpha)$ when $\rho = \rho_2$ and then to c when $\rho = c$. At the same time σ_1 increases from $-\infty$ to σ_{1M} and here we again have $\sigma_{1M} < -c$.

As ρ decreases from ρ_1 to ρ_2, we of course have Case (ii) above, and σ_0 decreases from ∞ to 0. A transition to Case (iii) takes place as ρ passes below ρ_2 and the point $(p(1), 0)$ bifurcates from the origin and $p(1)$ increases to $p_M = (1 - \alpha - cd(1 - \beta))/(\beta - \alpha)$ as $\rho \to c$. Observe that $p_M < 1$.

When $\rho \in [c, \rho_2)$ it is easily checked that the tangent to $\mathscr{C}$ through $(p(1), 0)$ has gradient

$$\sigma_3 = \frac{c(1 - \alpha - \rho d(1 - \beta))}{cd^2 \rho (1 - \beta) - (1 - \alpha)}.$$

The numerator is always positive but the denominator can take either sign and to understand what can happen we must distinguish two sub-cases. Observe that the denominator of σ_3 is zero when $\rho = \rho_3$ where

$$\rho_3 = \frac{1 - \alpha}{cd^2(1 - \beta)}.$$

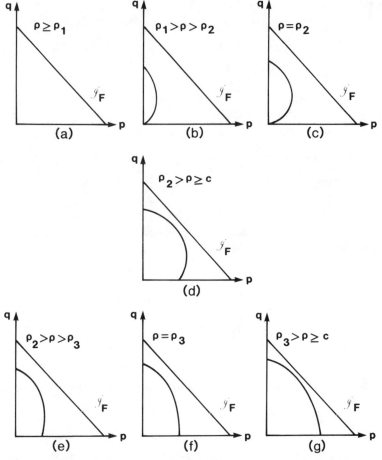

Figure 4.9. Model 1A: Growth of $\mathscr{C}$ as ρ decreases in Case II, where $1 < cd < (1 - \alpha)/(1 - \beta)$. (a), (b) and (c) show changes from Case (i) to Case (ii). (d) shows Case (iii) in subcase IIa, $c > \rho_3$. (e), (f) and (g) show the growth of $\mathscr{C}$ for Case (iii), subcase IIb. If $c = \rho_3$ then graph (f) is the final stage.

IIa. $\left(\dfrac{1 - \alpha}{1 - \beta}\right)^{1/2} < cd < \dfrac{1 - \alpha}{1 - \beta}$. For this case we have $\rho_3 < c$ and hence we always have $\sigma_3 < \infty$. Thus the bubble picture of $\mathscr{C}$ presented above for Case I still holds here when $\rho \geqslant \rho_2$, but with the changes we have noted, and as ρ decreases below ρ_2 the bottom end of the bubble moves along the p-axis. Figure 4.9 illustrates the sequence of changes.

IIb. $1 < cd \leqslant \left(\dfrac{1 - \alpha}{1 - \beta}\right)^{1/2}$. We now have $\rho_3 \geqslant c$ and hence ρ can attain ρ_3. As $\rho \to \rho_3+$, $\sigma_3 \to \infty$ and as ρ passes below ρ_3, σ_3 flips to a neighborhood of $-\infty$ and then increases through negative values as ρ decreases to c. The geometrical consequence of this discontinuous change in σ_3 is simply that the tangent of $\mathscr{C}$ through $(p(1), 0)$ becomes vertical when $\rho = \rho_3$ and as ρ passes below ρ_3 this tangent rotates in an anticlockwise direction. The evolution of $\mathscr{C}$ in Case IIb is also illustrated in Figure 4.9.

4.4   Convergence rates

In this section we resume our investigation of Model G under the conditions imposed in Section 2, namely, (4.1a) and $\mathscr{I}_G$ lying above $\mathscr{I}_F$ in $\mathscr{S}^0$. We know from (3.13) and Theorem 4.1 that the boundary equilibrium point $(0, \rho)$ is GAS but the proof of this fact does not indicate whether orbits starting in $\mathscr{R}_3^0$ must enter $\mathscr{R}_2^0$ in order to reach $(0, \rho)$. The following theorem resolves this question under very mild assumptions.

Theorem 4.2.

Suppose that $q_F < \rho$, $F''(0+)$ and $g = G'(1-)$ are finite, and let $\Delta = (1 - \alpha)q_F/\rho + \alpha$. Then $p_n \to 0$ geometrically fast, as given by (4.18) below. Moreover $\mathscr{L} = \lim_{n\to\infty} (\rho - q_n)/p_n$ exists and the following three cases can occur:

(a) $\mathscr{L} = g\dfrac{1 - \beta}{\Delta - \beta}$ *if $\beta < \Delta$;*

(b) $\mathscr{L} = \infty$ *if $\beta = \Delta$; and*

(c) $\mathscr{L} = \pm\infty$ *if $\beta > \Delta$ and the sign is determined by the value of (p_1, q_1).*

Remarks. A similar result holds when $\mathscr{I}_F$ lies above $\mathscr{I}_G$ in $\mathscr{S}^0$, in which case $(1, 0)$ is the GAS equilibrium. Specifically, let $p_G < 1$, $G''(0+) < \infty$, $f = F'(1-) < \infty$ and let $D = (1 - \beta)p_G + \beta$. Then $q_n \sim$ const. D^n and $\mathscr{M} = \lim_{n\to\infty} (1 - p_n)/q_n$ exists. The following cases can occur: $\mathscr{M} = f(1 - \alpha)/(D - \alpha)$ if $\alpha < D$; $\mathscr{M} = \infty$ if $\alpha = D$; and $\mathscr{M} = \pm\infty$ if $\alpha > D$.

Proof. First observe that $\Delta < 1$. The theorem is proved by determining the rates at which the sequences $\{p_n\}$ and $\{\rho - q_n\}$ converge to zero.

Since $q_F < \rho$ we can choose $\varepsilon > 0$ such that $q_F < \rho - \varepsilon$ and

$$\Delta_\varepsilon = (1 - \alpha)\frac{q_F + \varepsilon}{\rho - \varepsilon} + \alpha < 1.$$

There exists n' such that $q_n > \rho - \varepsilon$ iff $n \geqslant n'$ and hence from (4.2a) we obtain, by increasing n' if necessary,

$$p_{n+1} \leqslant (1 - \alpha)F\left(\frac{p_n}{\rho - \varepsilon}\right) + \alpha p_n \leqslant \Delta_\varepsilon p_n \quad (n \geqslant n').$$

Iteration of this inequality yields $p_{n+1} \leqslant p_{n'}\Delta_\varepsilon^{n-n'}$ and in particular

$$\sum_{n=1}^{\infty} p_n < \infty. \tag{4.15}$$

Let $\delta_n = \rho - q_n$ and $d_n = |\delta_n|$. Then (4.2b) can be recast as

$$\delta_{n+1} = \beta\delta_n + \gamma_n \tag{4.16}$$

where

$$\gamma_n = \rho(1 - \beta)[1 - G(q_n/(p_n + q_n))] \leqslant \rho(1 - \beta)gp_n/(p_n + q_n)$$

and it follows from (4.15) that $\Sigma\,\gamma_n < \infty$. Iteration yields $d_{n+1} \leqslant d_1\beta^n + \Sigma_{m=1}^{n}\gamma_{n-m}\beta^m$ and hence

$$\sum_{n=1}^{\infty} d_{n+1} \leqslant \beta\left[d_1 + (1 - \beta)^{-1}\sum_{n=1}^{\infty}\gamma_n\right] < \infty.$$

Now write (4.2a) as

$$p_{n+1} = p_n(\Delta + \varepsilon_n) \tag{4.17}$$

where $\varepsilon_n = p_{n+1}/p_n - \Delta$. Taylor's theorem yields

$$\varepsilon_n = (1 - \alpha)\left[\frac{\rho - q_n}{(p_n + q_n)(p_n + \rho)}F'(\zeta_n') - \frac{p_n q_F}{p_n + \rho} + \frac{p_n}{(p_n + \rho)^2}F''(\zeta_n'')\right]$$

where ζ_n' and ζ_n'' are $O(p_n)$. It follows that $\Sigma|\varepsilon_n| < \infty$ and iteration of (4.17) yields

$$p_n = p_1\Delta^{n-1}\prod_{i=1}^{n-1}(1 + \Delta^{-1}\varepsilon_i) \sim p_1\Gamma\Delta^n \quad (n \to \infty) \tag{4.18}$$

where

$$0 < \Gamma = \Delta^{-1}\prod_{i=1}^{\infty}(1 + \Delta^{-1}\varepsilon_i) < \infty.$$

To see that Γ really is positive note that $\Gamma = 0$ only if $\varepsilon_i = -\Delta$ for some i and this occurs only if $p_{i+1} = 0$ which is impossible since we only consider orbits starting within $\mathscr{S}^0$. Thus (4.18) gives the exact rate at which $\{p_n\}$ converges to zero.

We turn our attention now to $\{\delta_n\}$. Iteration of (4.16) yields

$$\delta_{n+1} = \delta_1 \beta^n + \sum_{m=0}^{n-1} \gamma_{n-m} \beta^m$$

and since $1 - G(q_n/(p_n + q_n)) \sim p_n g/\rho$, (4.18) shows that $\gamma_n \sim K\Delta^n$ where $K = (1 - \beta)p_1 \Gamma g$.

Suppose that $\beta < \Delta$. Then

$$\sum_{m=0}^{n-1} \gamma_{n-m} \beta^m = \Delta^n \sum_{m=0}^{n-1} (\gamma_{n-m}/\Delta^{n-m})(\beta/\Delta)^m$$

$$\sim \Delta^n K \sum_{m=0}^{\infty} (\beta/\Delta)^m, \quad (n \to \infty)$$

and we have used the dominated convergence theorem in taking the limit. It follows that

$$\delta_n \sim K\Delta^n/(\Delta - \beta)$$

and the assertion in (a) now follows from (4.18).

When $\Delta = \beta$ we have $\beta^{-n}\gamma_n \to K$ whence

$$\sum_{m=0}^{n-1} \gamma_{n-m} \beta^m = (n\beta^n)\left(n^{-1} \sum_{m=1}^{n} \gamma_m \beta^{-m} \right) \sim Kn\beta^n$$

and $\delta_n \sim Kn\beta^{n-1}$. It follows now from (4.18) that

$$(\rho - q_n)/p_n \sim (\beta^{-1} - 1)gn \quad (n \to \infty),$$

which gives assertion (b).

Finally, if $\Delta < \beta$ the series $\Sigma_{m=1}^{\infty} \gamma_m \beta^{-m}$ is convergent, whence

$$\lim_{n \to \infty} \beta^{-n}\delta_n = \tau = \beta^{-1}\left(\delta_1 + \sum_{m=1}^{\infty} \gamma_m \beta^{-m} \right).$$

It follows that $|\rho - q_n|/p_n \to \infty$ and that the sign of $(\rho - q_n)/p_n$ when n is large is the same as that of τ. We shall show later for Model 1A that τ can take either sign, but apart from this the proof is complete.

Restricting attention now to Model 1A, the derivative conditions required for Theorem 4.2 are satisfied, $g = k_{21}^{-1}$ and

$$\Delta = UV^{-1}k_{12}(1 - \beta) + \alpha.$$

Each of the three cases in Theorem 4.2 can occur. To see this let $V = Uk_{12}$, $\beta > 1/2$ and $\alpha = 2\beta - 1$. This choice yields $\Delta = \beta$, $\rho = 2\beta k_{12}/(1 - \beta) > k_{12}$, and choosing k_{12} large enough ensures that $k_{21}\rho > 1$. These parameter values ensure that $(0, \rho)$ is GAS and case (b) obtains. Small variations of α will preserve the stability conditions but allow Δ to vary either side of β.

To see that τ can take either sign when $\Delta < \beta$ it is sufficient to examine the de Wit case $k_{12}k_{21} = 1$. In that case the definition of γ_n and Equation (3.7) yield

$$\gamma_n = \rho(1 - \beta) - (q_{n+1} - \beta q_n) = \rho(1 - \beta)(1 - q_n/(k_{12}p_n + q_n))$$

$$= (V/U)(p_{n+1} - \alpha p_n)$$

whence

$$\tau = \delta_1/\beta + (V/U)((1 - \alpha/\beta)\Omega - p/\beta)$$

where $\Omega = \Sigma_{n=1}^{\infty} p_n \beta^{-n} < \infty$. If $q_1 < \rho$ then the orbit starting from (p_1, q_1) must enter $\mathscr{R}_2$ whence $\tau > 0$. We will now show that if $p_1 < \beta\rho$ and q_1 is sufficiently large then τ is negative. This will follow if Ω can be bounded above by a quantity independent of q_1.

Consider an orbital point (p_n, q_n) in $\mathscr{R}_3^0$. Remembering that $k_{12}k_{21} = 1$ we have $q_n + k_{12}p_n = \rho + \varepsilon_n$, say, where $\varepsilon_n > 0$ (and is not the ε_n in the proof of Theorem 4.2) and Equation (3.7b) yields

$$q_{n+1} - (\rho - k_{12}p_n) = \frac{\varepsilon_n}{\rho + \varepsilon_n}(\beta\rho + \beta\varepsilon_n + k_{12}p_n(1 - \beta)) > 0.$$

Let $\bar{n} = \sup\{n : (p_n, q_n) \in \mathscr{R}_3^0\}$. If $\bar{n} < \infty$ we have $q_{\bar{n}+1} > \rho - k_{12}p_{\bar{n}} \geqslant \rho - k_{12}p_1$ and the last term is positive if p_1 is sufficiently small. But $(p_{\bar{n}+1}, q_{\bar{n}+1}) \in \mathscr{R}_2$ whence $q_{\bar{n}+1} = \min\{q_n\}$ and it follows that $q_n > \rho - k_{12}p_1 > 0$ for all n. This continues to be true if $\bar{n} = \infty$ for in this case $(p_n, q_n) \in \mathscr{R}_3^0$ for all n. Using this inequality in (3.7a) yields $p_{n+1} \leqslant \tilde{\Delta}p_n$ where $\tilde{\Delta} = \alpha + (1 - \alpha)k_{12}/(\rho - k_{12}p_1)$ and iteration of the last inequality yields $p_n \leqslant p_1\tilde{\Delta}^{n-1}$. Using the parameter values chosen above we obtain

$$\beta - \tilde{\Delta} = (1 - \beta)\left(1 - \frac{2(1 - \beta)}{2\beta - (1 - \beta)p_1}\right)$$

and since $\beta > 1/2$ the right-hand side is positive if p_1 is sufficiently small, that is, $\tilde{\Delta}/\beta < 1$. The required bound now follows, namely, $\Omega \leqslant p_1/(\beta - \tilde{\Delta})$.

We will now explore the consequences of Theorem 4.2 for Model 1A. When $\beta < \Delta$ assertion (a) of this theorem is

$$(\rho - q_n)/p_n \to \frac{1 - \beta}{k_{21}(\Delta - \beta)}$$

and using Equation (3.5) this result translates to the following result for the maximum seed pool number,

$$\left(\frac{M}{1 - \beta} - u_n\right)\Big/t_n \to \frac{X(1 - \beta)}{Yk_{21}(\Delta - \beta)}. \tag{4.19}$$

The isoclines $\mathscr{I}_F$ and $\mathscr{I}_G$ transform into t- and u-isoclines defined respectively by

$$u = k_{12} XY^{-1}\left(\frac{L}{1-\alpha} - t\right)\left(0 \leqslant t \leqslant \frac{L}{1-\alpha}\right)$$

and

$$u = \frac{M}{1-\beta} - Xk_{21}^{-1}Y^{-1}t\left(0 \leqslant t \leqslant \frac{Yk_{21}}{X} \cdot \frac{M}{1-\beta}\right).$$

Equation (4.19) shows that all orbits approach the GAS equilibrium at $(0, M(1-\beta)^{-1})$ along an asymptote which passes through this point and lies below the u-isocline in $\mathscr{S}^0$ and they make this approach geometrically fast. This is illustrated in Figure 4.10. For all the examples considered in Chapter 7 it is the case that $\beta < \Delta$ but in some examples Δ is very close to unity. In such a case the speed of approach to the equilibrium can be relatively slow, a situation which could give a false impression of long-term coexistence with one species ultimately having very low numbers. This occurs for two of the mixtures discussed in Chapter 7.

The case where $\beta = \Delta$ represents the limit obtained by letting Δ approach β from above and is manifested by rotating the approach-asymptote in a clockwise direction until it lies on the u-axis. This is illustrated by the lower orbits in Figure 4.11 which also illustrates the case where $\beta > \Delta$.

We now consider the case omitted in the statement of Theorem 4.2, that is, where $p_G = \rho$. The next theorem shows that all orbits approach

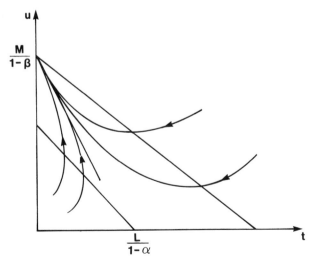

Figure 4.10. Model 1A: Typical final orbit configurations for maximum seed pool numbers when $\beta < \Delta$.

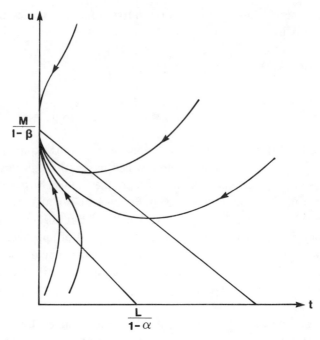

Figure 4.11. Model 1A: Typical final orbit configurations for maximum seed pool numbers when $\beta > \Delta$.

$(0, \rho)$ tangentially to $\mathscr{I}_G$ and that the speed of approach is far slower than in Theorem 4.2.

> *Theorem 4.3.*
> *Suppose $\mathscr{I}_G$ lies above $\mathscr{I}_F$ in $\mathscr{S}^0$, F is twice differentiable in a neighborhood of the origin, $q_F = \rho > 0$ and $g = G'(1-) < \infty$. Then $\lim_{n \to \infty} (\rho - q_n)/p_n = g$.*
> *Suppose also that $|F''(0+)| < \infty$ and define*
> $$D = (1 - \alpha)[\rho^{-1}(1 - g) - (2\rho^2)^{-1}F''(0+)].$$
> *Then $D \geqslant 0$ and if $D > 0$ we have*
> $$p_n \sim (Dn)^{-1} \quad \text{and} \quad \rho - q_n \sim g/Dn \quad \text{as} \quad n \to \infty.$$

Proof. Define ε_n by the equation $p_{n+1} = p_n(1 - \varepsilon_n)$. Using Taylor's expansion we obtain

$$\varepsilon_n = (1 - \alpha)\left(1 - p_n^{-1}F\left(\frac{p_n}{p_n + q_n}\right)\right)$$

$$= (1 - \alpha)\left(\frac{p_n - (\rho - q_n)}{p_n + q_n} - \frac{p_n}{2(p_n + q_n)^2}F''(\zeta_n)\right) \qquad (4.20)$$

where $0 < \zeta_n < p_n/(p_n + q_n) \to 0$, and $\varepsilon_n \to 0$ as $n \to \infty$. Ultimately $p_{n+1} < p_n$ whence $\varepsilon_n > 0$, and since $p_{n+1} = p_1 \Pi_{j=1}^n (1 - \varepsilon_j) \to 0$ we conclude that $\Sigma \varepsilon_n = \infty$.

The mean value theorem gives

$$G\left(\frac{q_n}{p_n + q_n}\right) = 1 - \frac{p_n}{p_n + q_n} G'(\xi_n), \quad \left(\frac{q_n}{p_n + q_n} < \xi_n < 1\right),$$

and defining $\gamma_n = (\rho - q_n)/p_n$, some algebraic manipulation yields the difference equation

$$\gamma_{n+1} = \beta c_n \gamma_n + J_n$$

where $c_n = (1 - \varepsilon_n)^{-1}$ and

$$J_n = \frac{\rho(1 - \beta)}{1 - \varepsilon_n} \cdot \frac{G'(\xi_n)}{p_n + q_n} \to (1 - \beta)g.$$

Iterating the difference equation yields

$$\gamma_{n+1} = \gamma_1 \beta^n \prod_{j=1}^n c_j + \sum_{m=0}^{n-1} J_{n-m} \beta^m \prod_{v=n-m+1}^n c_v.$$

The first product on the right-hand side can be written as $\Pi_{j=1}^n (\beta/(1 - \varepsilon_j))$ and this clearly tends to zero as $n \to \infty$. Choose b so that $\beta < b < 1$. The coefficient of J_{n-m} is bounded above by

$$K_1(\beta/b)^m \prod_{v=n-m+1}^n bc_v \leqslant K_2(\beta/b)^m,$$

since $b < 1 - \varepsilon_v$ for all except finitely many v, and the Ks are constants. It follows that the dominated convergence theorem is applicable, yielding

$$\lim_{n \to \infty} \gamma_n = \lim_{n \to \infty} J_n = \sum_{m=0}^\infty \beta^m = g,$$

the first assertion of the theorem.

Implicitly differentiating (4.2a) twice shows that the gradient of $\mathscr{I}_F$ at the point it intersects the q-axis is $F''(0+)/2F'(0+) - 1$, provided $F'(0+) > 0$, and similarly, that of $\mathscr{I}_G$ is $-g$. Thus the inequality $D \geqslant 0$ is a consequence of our hypothesis that $\mathscr{I}_G$ lies above $\mathscr{I}_F$ in $\mathscr{S}^0$ and that they meet at $(0, \rho)$. Our additional assumption $D > 0$ means that the isoclines do not osculate at $(0, \rho)$.

Returning to (4.20) and using the first assertion of the theorem, we obtain

$$\varepsilon_n \sim Dp_n \quad (n \to \infty) \tag{4.21}$$

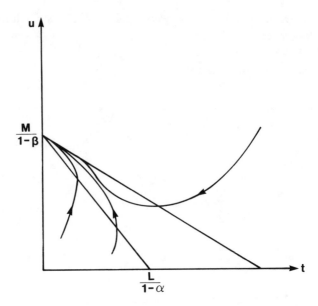

Figure 4.12. Model 1A: Typical final orbit configurations for maximum seed pool numbers when $\Delta = 1$.

Let $v_n = 1/p_n$ and rewrite the equation defining ε_n as $v_{n+1} - v_n = v_{n+1}\varepsilon_n$ whose solution is

$$v_n - v_1 = \sum_{m=1}^{n-1} \varepsilon_m/p_{m+1}.$$

Recalling that $p_{n+1}/p_n \to 1$, (4.21) implies that $\varepsilon_n/p_{n+1} \to D$ and it follows that $n^{-1}\sum_{m=1}^{n-1} \varepsilon_m/p_{m+1} \to D$. The final assertions of the theorem are now obvious.

If the conditions of Theorem 4.3 are satisfied and $g > 0$ then for all sufficiently large n we have $q_n < \rho$ and hence all orbits approach the GAS equilibrium within $\mathcal{R}_2$. The hypotheses are satisfied for Model 1A by taking $k_{12} = \rho > 1$ and then $g = k_{21}^{-1}$ and $D = (1 - \alpha)$ $(1 - (k_{21}\rho)^{-1}) > 0$. Equation (4.19) for the maximum seed pool numbers continues to hold but with $\Delta = 1$. Figure 4.12 illustrates orbit configurations for this case.

A case excluded by Theorem 4.3 is that where $D = 0$, that is, the isoclines meet tangentially at $(0, \rho)$. This situation cannot occur for Model 1A, given our assumption of no internal equilibria. For this reason we will not pursue the case $D = 0$ beyond remarking that if $|G''(1-)| < \infty$ then $p_n \sim (2Kn)^{-1/2}$ where $K = \rho^{-2}(1 - \alpha)$ $(1 - g - G''(1-)/2)$.

In concluding our discussion of Theorems 4.2 and 4.3 we mention once again that behavior of the other seed quantities can be inferred from the properties of $\{(t_n, u_n)\}$ described above. In particular, since $x_n \sim (1 - \alpha/\Delta)t_n$ and

$$M - y_n = \frac{M}{1 - \beta} - u_n - \beta\left(\frac{M}{1 - \beta} - u_{n-1}\right)$$

we find that

$$(M - y_n)/x_n \to M_2/(M_1 k_{12} k_{21}).$$

When $k_{12} k_{21} = 1$, as assumed by Rossiter et al. (1985), we see that in the long run $M - y_n$ is proportional to x_n and that the constant of proportionality is the ratio of the monoculture seed yields.

4.5 The case $\rho = 0$

Up till now we have assumed that $\rho > 0$ which certainly must be the case for Model 1A. For the sake of completing the theory we will now consider the degenerate case of Model G wherein $\rho = 0$. The system described by (3.6) then reduces to a one-dimensional system given by (3.6a) with $q_n = q_1 \beta^{n-1}$. We prove the following theorem. Recall that $z_n = p_n/(p_n + q_n)$.

Theorem 4.4.

If $\rho = 0$ and $0 < F'(0+) < \infty$ then $q_n \to 0$ and $p_n \to 1$. In addition: if $\alpha < \beta$ then $(1 - p_n)/q_n \to F'(0+)(1 - \alpha)/(\beta - \alpha)$; if $\alpha = \beta$ then $(1 - p_n)/nq_n \to (1 - \alpha)\beta^{-1}F'(0+)$; and if $\alpha > \beta$ then $\alpha^{-n}(1 - p_n) \to \alpha^{-1}[1 - p_1 + (1 - \alpha)\sum_{m=1}^{\infty} \alpha^{-m}(1 - F(z_m))]$.

Proof. The assertion of Theorem 3.2 continues to hold. Suppose that there is a positive number ε such that $p_n \geq \varepsilon$ for all n. Then $z_n \to 1$ as $n \to \infty$. Equation (3.6a) can be written as $\alpha^{-n-1}p_{n+1} - \alpha^{-n}p_n = (1 - \alpha)\alpha^{-n-1}F(z_n)$ and successive addition gives

$$p_n = p_1\alpha^{n-1} + (1 - \alpha)\alpha^n \sum_{m=1}^{n-1} \alpha^{-m-1}F(z_m)$$

$$= p_1\alpha^{n-1} + (1 - \alpha)\sum_{m=0}^{n-2} \alpha^m F(z_{n-m-1}). \tag{4.22}$$

The dominated convergence theorem may be used to show that $p_n \to 1$.

We show that $\{p_n\}$ is bounded away from zero in two stages. The first stage is to show that $\{p_n\}$ does not converge to zero. Suppose the contrary. It follows from (3.6a) that $F(z_n) \to 0$ whence $z_n \to 0$ and this implies that $p_n/q_n \to 0$. Since $q_{n+1}/q_n \to \beta$ the last conclusion and (3.6a) show that $q_n^{-1}F(z_n) \to 0$ and since $0 < F'(0+) < \infty$ it follows that $q_n^{-2}p_n \to 0$. Continuing this argument in an inductive manner we conclude that $\beta^{-nv}p_n \to 0$ for any positive integer v. However, (4.22) shows that $p_n > p_1\alpha^{n-1}$ and selecting v so that $\beta^{-v}\alpha > 1$ leads to the contradictory conclusion $\beta^{-nv}p_n \to \infty$.

It follows from the last argument that there is an increasing sequence of positive integers $\{n'\}$ such that $p_{n'} \geq \varepsilon > 0$. Equation (3.6a) then yields

$$p_{n'+1} \geq (1 - \alpha)F\left(\frac{\varepsilon}{\varepsilon + q_{n'}}\right) + \alpha\varepsilon.$$

Now, if v is large enough then

$$(1 - \alpha)F\left(\frac{\varepsilon}{\varepsilon + q_v}\right) + \alpha\varepsilon > \varepsilon$$

and we conclude that $p_{n'+1} \geq \varepsilon$ if $n' \geq v$. Repeating this argument we conclude that $p_{n'+m} \geq \varepsilon$ if $n' \geq v$ and m is any positive integer, that is, $\{p_n\}$ is bounded away from zero.

If $\phi_n = 1 - p_n$ then (4.22) can be written as

$$\phi_n = (1 - p_1)\alpha^{n-1} + (1 - \alpha)\sum_{m=0}^{n-2} \alpha^m(1 - F(z_{n-m-1})).$$

As $n \to \infty$, $1 - F(z_{n-m-1}) \sim F'(0+)q_{n-m-1} = F'(0+)q_1\beta^{n-m-2}$ and the first two of the convergence rate assertions follow immediately. For the third, rewrite the last equation as

$$\alpha^{-n}\phi_n = \alpha^{-1}\left[1 - p_1 + (1 - \alpha)\sum_{m=1}^{n-1} \alpha^{-m}(1 - F(z_m))\right]$$

and observe that $\sum_{m=1}^{\infty} \alpha^{-m}(1 - F(z_m)) < \infty$.

5

Analysis of Model G – one interior equilibrium

5.1 Introduction

In this chapter we shall determine the asymptotic behavior of Model G when it is constrained so as to have exactly one internal equilibrium $\varepsilon = (\bar{p}, \bar{q}) \in \mathscr{S}^0$. This can occur in three ways. First, imagine that the isoclines in Figure 4.1 are deformed so that they touch at ε, but without crossing in $\mathscr{S}^0$; see Figure 5.1. Thus $\mathscr{R}_2$ in Figure 4.1 is deformed so that its interior has two components and the boundaries of these components have the single point ε in common. We can partition $\mathscr{R}_2$ as $\mathscr{R}_2^q \cup \{\varepsilon\} \cup \mathscr{R}_2^p$, where the definition of $\mathscr{R}_2^q$ and $\mathscr{R}_2^p$ should be clear from Figure 5.1. This is a situation that cannot arise for Model 1A and hence we shall not explore it in any detail. We simply remark that the positive invariance assertions of Theorem 4.1 are still valid and hence an orbit starting in $\mathscr{R}_1$ either enters $\mathscr{R}_2^q$ and then converges to $(0, \rho)$, or it converges to ε either after entering $\mathscr{R}_2^q$ or without entering $\mathscr{R}_2$ at all. Similar remarks apply to orbits starting in $\mathscr{R}_3$.

A second realisation of a single interior equilibrium occurs when $\mathscr{I}_F$ lies above $\mathscr{I}_G$ near the q-axis, but the reverse is true near the p-axis. This is illustrated in Figure 5.2. The orbital drift directions shown by the arrows in this figure can be inferred from the discussion in Section 4.1, (following (4.2)) and they suggest that ε is GAS. By using the same ideas employed to prove Theorem 4.1 we will show in the next section that this is the case (Theorem 5.2). The hypotheses of this theorem constrain the shape of the isoclines near ε. In fact they must be non-decreasing in a neighborhood of the equilibrium point because otherwise it would not be possible to represent $\mathscr{I}_G$ (respectively $\mathscr{I}_F$) as a function of q (respectively p). We will demonstrate this in Theorem 5.3 in preparation for our discussion of convergence rates and the asymptotic direction of approach to ε.

A traditional approach to proving global asymptotic stability of an internal equilibrium is seeking a suitable Liapunov function. In Section

79

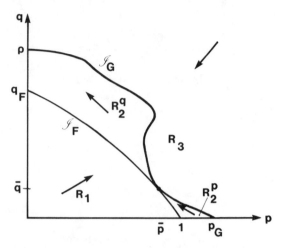

Figure 5.1. Model G: One interior equilibrium is generated by the isoclines touching without crossing.

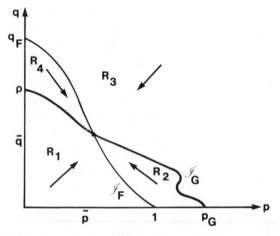

Figure 5.2. Model G: The isoclines cross once to give a single interior (stable) equilibrium.

3 we will pay our respects to this tradition by exhibiting a Liapunov function for Model 1A.

The third way in which an interior equilibrium can occur is when $\mathscr{I}_G$ lies above $\mathscr{I}_F$ near the q-axis and the reverse is true near the p-axis; see Figure 5.3. The orbital drift directions shown by the arrows suggest that orbits converge to one or other of the boundary equilibria, the choice depending on the starting point. In Section 5 we shall show that ε cannot be AS and prove some other results which support this intuition. Except when $\alpha = \beta$ it does not seem possible to determine domains of

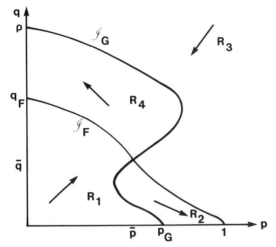

Figure 5.3. Model G: Isoclines cross once to give a single interior (unstable) equilibrium.

attraction of the boundary equilibria, but for Model 1A it is possible to obtain inner estimates of them by using Liapunov functions.

There is one final case that can occur for Model 1A, namely, the degenerate case wherein the isoclines coincide. In this case every point on the isocline is an equilibrium point. We shall examine this situation for Model G in Section 6 and show that each orbit converges to some point on the isocline.

Finally, we remark that it seems the only cases of Model 1A relevant to the experimental work of Rossiter et al. (1985) are the case of no internal equilibria or a single GAS equilibrium.

5.2   Stable interior equilibrium

In this section we consider the situation illustrated in Figure 5.2. Let $\bar{z} = \bar{p}/(\bar{p} + \bar{q})$ and observe that $0 < \bar{z} < 1$. If $z < \bar{z}$ then R_z meets $\mathscr{I}_G$ before it meets $\mathscr{I}_F$ and the discussion leading to (4.3) shows that this occurs iff

$$F(z) > \frac{z}{1-z}\rho G(1-z) \quad (0 < z < \bar{z}). \tag{5.1}$$

This condition is the mathematical representation of our intuitive idea of $\mathscr{I}_F$ lying above $\mathscr{I}_G$ to the left of the vertical line through $(\bar{p}, \bar{q})$. Similarly the notion that $\mathscr{I}_F$ lies below $\mathscr{I}_G$ to the right of this line is represented by the condition

$$F(z) < \frac{z}{1-z}\rho G(1-z) \quad (\bar{z} < z < 1). \tag{5.2}$$

The following lemma is a useful consequence of these considerations and it will be used in Section 4.

Lemma 5.1.

Let $\mathcal{O}$ be the closed rectangle with opposite corners $(0, 0)$ and $(\bar{p}, \bar{q})$. Then $\mathcal{I}_F$ and $\mathcal{I}_G$ do not intersect the interior of $\mathcal{O}$. Moreover if $F(z) > F(\bar{z})$ and $G(1 - z) < G(1 - \bar{z})$ whenever $z > \bar{z}$ then $\mathcal{I}_F$ and $\mathcal{I}_G$ intersect $\mathcal{O}$ in the single point $(\bar{p}, \bar{q})$.

Proof. Construct $\mathcal{I}_F$ as follows. Choose $z \in [0, 1]$, $p(z) = F(z)$ and then $q(z) = (1 - z)p(z)/z$. Then $\mathcal{I}_F$ consists of the points $((p(z), q(z))$ generated by allowing z to range through $[0, 1]$. A similar construction gives $\mathcal{I}_G$. Now consider that segment of $\mathcal{I}_F$ corresponding to $\bar{z} \leqslant z \leqslant 1$. Since $p(z) \geqslant \bar{p}$, $(p(z), q(z)) \notin \mathcal{O}^0$, and since $(p(z), q(z))$ is nearer to $(0, 0)$ than any point in $R_z \cap \mathcal{I}_G$, no point on $\mathcal{I}_G$ corresponding to $z > \bar{z}$ lies in $\mathcal{O}$.

If $F(z) = F(\bar{z})$ for some $z > \bar{z}$ then $p(z) = \bar{p}$ but $q(z) < \bar{q}$ and hence a portion of $\mathcal{I}_F$ will run down part of the vertical edge of $\mathcal{O}$ through ε. Conversely, if $F(z) > F(\bar{z})$ when $z > \bar{z}$ then $(p(z), q(z)) \notin \mathcal{O}$ if $z > \bar{z}$. The proof is completed by applying similar considerations to the segment of $\mathcal{I}_G$ corresponding to $z < \bar{z}$.

The construction of the isoclines used in this proof shows that for each $z \in (0, 1)$, each isocline intersects R_z exactly once. This imposes restrictions on the extent to which $i(\cdot)$ and $j(\cdot)$ can oscillate. We now sketch the proof of the following fundamental result.

Theorem 5.2.

Suppose $\rho > 0$ and (5.1) and (5.2) are valid. Then $\mathcal{R}_2$, $\mathcal{R}_4$, $\mathcal{R}_1 \cup \mathcal{R}_4$ and $\mathcal{R}_3 \cup \mathcal{R}_2 \cup \mathcal{R}_4$ are positively invariant. In addition $(\bar{p}, \bar{q})$ is GAS.

Proof. The proof is very similar to that of Theorem 4.1. For example, if $(p, q) \in \mathcal{R}_4 \cap \mathcal{S}^0$ then $p \leqslant F(z)$ and $q \geqslant \rho G(1 - z)$ $(0 < z < \bar{z})$ and at least one inequality is strict. Then (see Equation (4.4))

$$\sigma(p, q) = (1 - \alpha)(F(z) - p) + (1 - \beta)(p/q)(q - \rho G(1 - z)) > 0$$

and hence $z' > z$. It follows that $p' \leqslant F(z')$ and $q' \geqslant \rho G(1 - z')$; see the argument starting at (4.5). Thus $\mathcal{R}_4$ is positively invariant. Similarly if $(p, q) \in \mathcal{R}_2 \cap \mathcal{S}^0$ then $z' < z$ and $\mathcal{R}_2$ is invariant also.

The spiraling directions of trajectory segments in $\mathcal{R}_1$ and $\mathcal{R}_3$ depend on the relation between α and β. As in the proof of Theorem 4.1 we will assume $\alpha \geqslant \beta$.

If $\alpha = \beta$ then

$$\sigma(p, q) = (1 - \alpha)\left[F(z) - \frac{\rho z}{1 - z}G(1 - z)\right]$$

and hence $z' > z$ if $z < \bar{z}$ and $z' < z$ if $z > \bar{z}$. Thus trajectories always spiral toward $R_{\bar{z}}$. On $R_{\bar{z}}$, $z' = \bar{z}$ and hence a trajectory starting on $R_{\bar{z}}$ moves along it toward $(\bar{p}, \bar{q})$. Suppose $(p, q) \in \mathcal{R}_3$. Then $p > F(z)$ and if $z' < z$ we have

$$p' - F(z') = F(z) - F(z') - \alpha(p - F(z)) > 0$$

and hence $(p', q') \notin \mathcal{R}_1 \cup \mathcal{R}_4$. Similarly, if $z' > z$ then $q' > \rho G(1 - z')$, and we conclude that $\mathcal{R}_3 \cup \mathcal{R}_2 \cup \mathcal{R}_4$ is invariant, as is $\mathcal{R}_1 \cup \mathcal{R}_2 \cup \mathcal{R}_4$. Very similar arguments apply when $\alpha \neq \beta$. We can now argue as in the proof of Theorem 4.1 to show that any trajectory starting in $\mathcal{S}^0$ converges to $(\bar{p}, \bar{q})$. This concludes the proof.

When $\alpha = \beta$ we have seen above that the infinite sectors in $\mathcal{S}^0$ whose common boundary is $R_{\bar{z}}$ are $\mathcal{R}_p$ and $\mathcal{R}_q$, with $\mathcal{R}_p$ lying above $\mathcal{R}_q$. Now consider the case $\alpha > \beta$. The case $\alpha < \beta$ is obtained by interchanging the roles of p and q. As in Section 4.2 let $\mathcal{C} = \mathcal{R}_p \cap \mathcal{R}_q$ be the curve given by $\sigma(p, q) = 0$. It follows from (4.6) that $\mathcal{C}$ is bounded. Moreover if $z \geqslant \bar{z}$ then $\mathcal{C}$ can be determined as in the discussion following Equation (4.6). In particular $(\bar{p}, \bar{q}) \in \mathcal{C}$, $\mathcal{C}$ intersects each ray R_z in exactly one point, and $\mathcal{C}$ hits the p axis at the point $(p(1), 0)$ where $p(1)$ is given by Equation (4.7b).

The situation is less clear when $z < \bar{z}$ because $p(z)$ may be negative if z is sufficiently small. Suppose $(p, q) \in \mathcal{R}_1$ and just below $\mathcal{I}_G$. More specifically, fix $p < p'$ and define $\eta > 0$ by $q = \rho G(1 - z) - \eta$. Then

$$\sigma(p, q) = (1 - \alpha)(F(z) - p) - (1 - \beta)\frac{z}{1 - z}\eta,$$

and the second term $\to 0$ as (p, q) approaches $\mathcal{I}_G$ along R_z while the first term, which is positive, remains bounded away from zero. It follows that $z' > z$ if (p, q) is close to $\mathcal{I}_G$.

Now suppose $z > \bar{z}$. Then (5.2) holds whence

$$\sigma(p, q) < (\alpha - \beta)(p - F(z)) < 0$$

and $z' < z$. It must be the case that $\mathcal{C}$ extends through $\mathcal{R}_1$ and lies above $R_{\bar{z}}$. Let

$$H(z) = (1 - \beta)\frac{\rho z}{1 - z}G(1 - z) - (1 - \alpha)F(z).$$

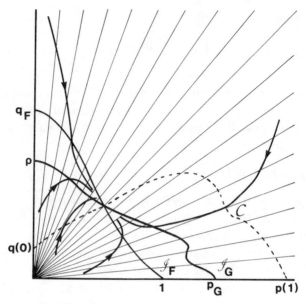

Figure 5.4. A possible configuration of $\mathscr{C}$ when $z_0 = 0$.

Now $H(\bar{z}) > 0$ and, by continuity, $H(z) > 0$ if $z_0 < z < \bar{z}$ where z_0 is the largest root in $[0, \bar{z})$ of the equation $H(z) = 0$. If $z \in (z_0, \bar{z})$ we can solve the equation $\sigma(p, q) = 0$ for $p = p(z)$ and then $q = q(z) = (1 - z)p(z)/z$. Thus in $\mathscr{R}_1$, $\mathscr{C}$ is a curve which is intersected exactly once by each R_z with $z_0 < z < \bar{z}$. There are three possible cases:

(i) The root $z_0 > 0$ and $\mathscr{C}$ terminates at the origin tangentially to R_{z_0}.

(ii) The transition case where $z_0 = 0$ but $\mathscr{C}$ still terminates at the origin tangentially with the q-axis.

(iii) We still have $z_0 = 0$ and $\mathscr{C}$ terminates at the point $(0, q(0))$ where $q(0)$ is given by (4.7a), but now $q(0) < \rho$.

A possible configuration of orbits is illustrated in Figure 5.4 for case (iii).

Let us consider Model 1A. The conditions (5.1) and (5.2) ensuring a GAS interior equilibrium ε reduce in this case to

$$c > \rho \quad \text{and} \quad \rho d > 1 \quad (\text{whence } cd > 1), \tag{5.3}$$

and ε is given by

$$\bar{p} = \frac{d(c - \rho)}{cd - 1} \quad \text{and} \quad \bar{q} = \frac{c(\rho d - 1)}{cd - 1}. \tag{5.4}$$

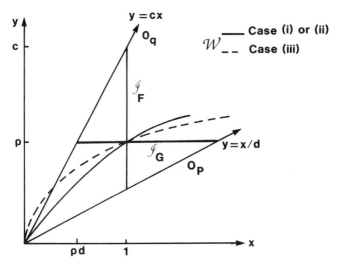

Figure 5.5. Model 1A: The transformed state space when $c > \rho$ and $\rho d > 1$.

Since (5.3) implies that c or $d > 1$, we see that a necessary condition for $(\bar{p}, \bar{q})$ to be GAS is that at least one of the strains gains by competition. To understand the configuration of $\mathcal{C}$ it is easiest to work in the transformed state space $\mathcal{W}$ introduced in Section 4.3. Then $\mathcal{C}$ is represented by the hyperbola

$$y = \phi(x) = \frac{(1 - \beta)\rho x}{1 - \alpha + (\alpha - \beta)x},$$

shown in Figure 5.5.

Since $\phi'(0+) = ((1 - \beta)/(1 - \alpha))\rho$, Cases (i) and (ii) occur iff

$$\frac{1 - \beta}{1 - \alpha}\rho \leqslant c \tag{5.5}$$

with Case (i) corresponding to strict inequality. The equation $H(z) = 0$ reduces to a linear equation whose solution is

$$z_0 = \frac{c \ \rho d - (1 - \alpha)/(1 - \beta)}{d((1 - \alpha)/(1 - \beta))c - \rho}$$

and which is the slope of the tangent to $\mathcal{C}$ at the origin. The point of intersection, $(p(1), 0)$, of $\mathcal{C}$ with the p-axis is given by Equation (4.14).

Suppose α is fixed and that β decreases from just below α. Then, because of (5.3), $p(1)$ decreases from large values and z_0 increases. If $\rho/(1 - \alpha) > c$ then as β decreases the inequality (5.5) reverses and the end point of $\mathcal{C}$ at $(0, 0)$ moves out from the origin along the q-axis at the point $(0, q(0))$ where $q(0)$ is given by (4.14).

5.3 A Liapunov function for Model 1A

Suppose conditions (5.3) are satisfied for Model 1A, let

$$V(\mathbf{x}) = (1 - \alpha)^{-1}[p - \bar{p} - \bar{p}\log(p/\bar{p})]$$
$$+ (c(1 - \beta))^{-1}[q - \bar{q} - \bar{q}\log(q/\bar{q})]$$

and observe that $V(\mathbf{x}) \to \infty$ as $\|\mathbf{x}\| \to \infty$. We now show that $\dot{V}(\mathbf{x}) < 0$ if $\mathbf{x} \neq \varepsilon$.

Some algebraic manipulation yields

$$\dot{V}(\mathbf{x}) = p\left(\frac{c}{cp + q} - 1\right) - \frac{\bar{p}}{1 - \alpha}\log(\alpha + (1 - \alpha)c/(cp + q))$$

$$+ \frac{pq}{c}\left(\frac{d}{p + dq} - 1\right) - \frac{\bar{q}}{c(1 - \beta)}\log(\beta + (1 - \beta)\rho d/(p + dq)).$$

Once again we change to (x, y) co-ordinates in $\mathscr{W}$ and examine $\dot{V}$ along the rays $y = mx$ where $d^{-1} \leqslant m \leqslant c$. Defining $f(x) = (cd - 1)$ $\dot{V}(p(x, mx), q(x, mx))$, where $(p(x, y), q(x, y))$ is the inverse image of $(x, y) \in \mathscr{W}$ induced by the map (4.10), we have

$$f(x) = d(c - m)(1 - x) - (md - 1)(\rho m^{-1} - x)$$
$$- \frac{d(c - \rho)}{1 - \alpha}\log(\alpha + (1 - \alpha)/x)$$
$$- \frac{\rho d - 1}{1 - \beta}\log(\beta + (1 - \beta)(\rho/mx)). \tag{5.6}$$

We will show that $f(x) < 0$ if $0 \leqslant x < \infty$ and $d^{-1} \leqslant m \leqslant c$, unless $x = 1$ and $m = \rho$, in which case $f(1) = 0$.

First observe that $f(0+) = -\infty$. Differentiation yields

$$f'(x) = 1 - cd + \frac{d(c - \rho)}{x(1 - \alpha + \alpha x)} + \frac{\rho(d\rho - 1)}{x(\rho(1 - \beta) + \beta mx)}$$

which is decreasing, $f'(0+) = \infty$ and $f'(\infty) = 1 - cd < 0$. It follows that $f(x)$ has a single maximum. Further algebra shows that

$$f'(1) = \frac{\beta(d\rho - 1)(\rho - m)}{\rho(1 - \beta) + \beta m} \begin{cases} >0 & \text{if} \quad d^{-1} \leqslant m < \rho \\ =0 & \text{if} \quad m = \rho \\ <0 & \text{if} \quad \rho < m \leqslant c. \end{cases}$$

With rather more manipulation we find that

$$f'(\rho/m) = \frac{[(cd - 1)(1 + \alpha\rho/m) - \alpha(\rho d - 1)][m - \rho]}{\rho(1 - \alpha + \alpha\rho/m)}.$$

The first term in square brackets exceeds

$$(cd - 1)(1 + \alpha\rho/m) - \alpha(cd - 1) > 0$$

whence

$$f'(\rho/m) \begin{cases} <0 & \text{if} \quad d^{-1} \leqslant m < \rho \\ =0 & \text{if} \quad m = \rho \\ >0 & \text{if} \quad \rho < m \leqslant c. \end{cases}$$

We deduce from these inequalities that the maximum of $f(x)$ lies strictly between unity and ρ/m when these differ, and it lies at unity when $m = \rho$.

The arguments of the logarithms at (5.6) can be written in the form $1 - AB$ where A is $1 - \alpha$ or $1 - \beta$, and hence is in $(0, 1)$, and $B < 1$. The inequality

$$-\log(1 - AB) \leqslant \frac{AB}{1 - B} \tag{5.7}$$

yields

$$f(x) \leqslant \rho^{-1}(m - \rho)[\rho(m^{-1} - d) + x(\rho d - 1)] \equiv \lambda(x).$$

When $m \neq \rho$ it is easily seen that the linear function $\lambda(x)$ satisfies $\lambda(1)$, $\lambda(\rho/m) < 0$ and hence $\lambda(x) < 0$ if x lies between unity and ρ/m. It follows that $f(x) < 0$ for all $x \geqslant 0$. If $m = \rho$ then $f(1) = 0$ and $f(x) < f(1)$ if $x \neq 0$. Thus $\dot{V}(\mathbf{x}) < 0$ if $\mathbf{x} \neq \boldsymbol{\varepsilon}$ and it follows that $\boldsymbol{\varepsilon}$ is GAS.

5.4   Convergence rates

We now examine how trajectories approach the GAS interior equilibrium $\boldsymbol{\varepsilon}$ and in particular we look at the question of the direction of final approach to $\boldsymbol{\varepsilon}$. Our main theorem requires some preliminary results about the configuration of the isoclines near $\boldsymbol{\varepsilon}$ and the eigenstructure of the Jacobian matrix $\mathscr{J}(\boldsymbol{\varepsilon})$ (see Equation (3.14)).

Theorem 5.3.
Suppose that both (5.1) and (5.2) are satisfied. Then
    (i) $i(\cdot)$ [respectively $j(\cdot)$] is decreasing in an open interval containing $\bar{p}$ [respectively $\bar{q}/\rho$]. If $i'(\bar{p}) < 0$ then $j'(\bar{q}/\rho) > -\infty$ and

$$i'(\bar{p})j'(\bar{q}/\rho) \geqslant 1. \tag{5.8}$$

If $i'(\bar{p}) = 0$ then $j'(\bar{q}/\rho) = -\infty$. In either case the matrix $\mathscr{J}(\boldsymbol{\varepsilon})$ has real distinct eigenvalues λ_1, λ_2 satisfying $0 < \lambda_2 \leqslant \lambda_1 \leqslant 1$.
    (ii) Suppose now that $i'(\bar{p})$ and $j'(\bar{q}/\rho)$ are finite and that strict inequality holds at (5.8). Then $0 < \lambda_2 < \lambda_1 < 1$ and λ_1 and λ_2 have eigenvectors of the form $\mathbf{e}_1 = (1, -u)^T$ and $\mathbf{e}_2 = (v, 1)^T$, respectively, where $u > 0$ and $v > 0$. Finally, u satisfies

$$-1/j'(\bar{q}/\rho) < u < -i'(\bar{p}). \tag{5.9}$$

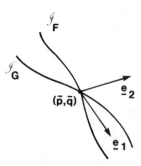

Figure 5.6. Model G: Configuration of the eigenvectors of $\mathscr{J}(\bar{p}, \bar{q})$ relative to the isoclines.

Remarks. (i) As the proof will reveal, (5.8) is strict if the isoclines intersect without osculating, as in the case of Model 1A, and equality holds otherwise.

(ii) Inequality (5.9) implies that e_1 lies between the south-easterly pointing tangential rays of the isoclines at $(\bar{p}, \bar{q})$. It follows that e_2 points into $\mathscr{R}_3$, as illustrated in Figure 5.6.

Proof. (i) Our assumptions are that $\mathscr{I}_F$ lies strictly above $\mathscr{I}_G$ in $(0, \bar{p})$ and the reverse is true in $(\bar{p}, 1)$. Suppose that $i'(\bar{p}) > 0$. Then $\mathscr{I}_G$ is forced to move away from Op in an interval $(\bar{p}, \bar{p} + \eta)$, as illustrated in Figure 5.7. More precisely, given $\eta > 0$ there is a point (p', q') on $\mathscr{I}_G$ satisfying $\bar{p} < p' < \bar{p} + \eta$ and $q' > \bar{q}$. Since $\mathscr{I}_G$ terminates at $(0, p_G)$, the intermediate value theorem shows that there exists $p'' > \bar{p} + \eta$ such that $(p'', \bar{q})$ lies on $\mathscr{I}_G$. However, this contradicts the fact that $\mathscr{I}_G$ can be specified as a function of q; see (3.18). We conclude that $i'(\bar{p}) \leqslant 0$. If $i'(\bar{p}) = 0$ it is clear that $j'(\bar{q}/\rho) = -\infty$. A similar argument applied to a left neighborhood of $\bar{p}$ shows that $j'(\bar{q}/\rho) \leqslant 0$.

We now show that the functional relationship (3.18) is strictly decreasing in an open neighborhood of $\bar{q}$. If this is not so then there is a sequence $\{q'_n\}$ such that $q'_n \uparrow \bar{q}$ and $p(q'_n) \geqslant \bar{p}$ and/or a sequence $\{q''_n\}$ such that $q''_n \downarrow \bar{q}$ and $p(q''_n) \geqslant \bar{p}$. Suppose the former is the case. For all n large enough the ray through $(p(q'_n), q'_n)$ intersects $\mathscr{I}_F$ at a point $(\tilde{p}_n, \tilde{q}_n)$ with $\tilde{p}_n < \bar{p}$ and $\tilde{q}_n < \bar{q}$. But this contradicts Lemma 5.1. A similar argument shows that $i(\cdot)$ is strictly decreasing in an open neighborhood of $\bar{p}$.

It follows that the functional relationship (3.18) is invertible in an open neighborhood of $\bar{q}$. More specifically, there is a function J defined in an open interval (p_1, p_2) containing $\bar{p}$ such that (3.18) can be re-expressed in the form

$$q = \rho J(p/\rho) \quad (p_1 < \bar{p} < p_2)$$

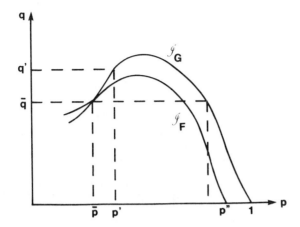

Figure 5.7. Model G: Configuration of the isoclines when $i'(\bar{p}) > 0$.

and J is decreasing and differentiable. If $p_2 - p_1$ is small enough then $i(\cdot)$ is also decreasing in (p_1, p_2) and since $\mathcal{I}_F$ crosses $\mathcal{I}_G$ from above as p increases, it follows that $i'(\bar{p}) \leqslant J'(\bar{p}/\rho)$. Since $j'(\bar{q}/\rho)J'(\bar{p}/\rho) = 1$, this inequality is equivalent to (5.8) – remember that $j'(\bar{q}/\rho) \leqslant 0$. In addition strict inequality in (5.8) holds iff $\mathcal{I}_F$ and $\mathcal{I}_G$ intersect without osculating.

Let
$$\mathcal{A} = (1 - \alpha)\bar{z}(1 - \bar{z})\bar{p}^{-1}F'(\bar{z}),$$
$$\mathcal{B} = (1 - \beta)\rho\bar{z}(1 - \bar{z})\bar{q}^{-1}G'(1 - \bar{z}),$$
$$2\mathcal{M} = \alpha + \beta + \mathcal{A} + \mathcal{B} \quad \text{and} \quad \mathcal{N} = \alpha\beta + \beta\mathcal{A} + \alpha\mathcal{B}.$$

It follows from (3.14) that the eigenvalues of $\mathcal{J}(\varepsilon)$ solve the equation $\lambda^2 - 2\mathcal{M}\lambda + \mathcal{N} = 0$, whose discriminant is

$$\mathcal{D} = 4(\mathcal{M}^2 - \mathcal{N}) = (\alpha - \beta + \mathcal{A} - \mathcal{B})^2 + 4\mathcal{A}\mathcal{B} \geqslant 0.$$

The eigenvalues are given by $\lambda_{1,2} = \mathcal{M} \pm (\mathcal{M}^2 - \mathcal{N})^{1/2}$ and satisfy $0 \leqslant \lambda_2 \leqslant \lambda_1$. Observe that $\mathcal{N} \leqslant \alpha + \beta - \alpha\beta < 1$, and, as we will now show, that $\mathcal{M} < 1$.

To see this, let ϕ be the inverse function of F. Then $F'(\bar{z})\phi'(\bar{p}) = 1$ and differentiation of (3.16) yields $i'(\bar{p}) = (1 - \bar{z})/\bar{z} - \bar{p}/\bar{z}^2F'(\bar{z})$. Since $\bar{p}/\bar{q} = \bar{z}/(1 - \bar{z})$ we obtain

$$\mathcal{A} = \frac{1 - \alpha}{1 - (\bar{p}/\bar{q})i'(\bar{p})} \leqslant 1 - \alpha. \tag{5.10}$$

Similarly

$$\mathcal{B} = \frac{1 - \beta}{1 - (\bar{q}/\bar{p})j'(\bar{q}/\rho)} \leqslant 1 - \beta. \tag{5.11}$$

Now $2(1 - \mathcal{M}) = (1 - \alpha - \mathcal{A}) + (1 - \beta - \mathcal{B})$ and $1 - \alpha - \mathcal{A} = 0$ iff $i'(\bar{p}) = 0$. This condition entails $j'(\bar{q}/\rho) = -\infty$, or $\mathcal{B} = 0$. Similarly $1 - \beta - \mathcal{B} = 0$ implies that $\mathcal{A} = 0$, and we conclude that $\mathcal{M} < 1$ in both cases.

Now consider

$$1 - (2\mathcal{M} - \mathcal{N}) = (1 - \alpha)(1 - \beta) - \mathcal{A}(1 - \beta) - \mathcal{B}(1 - \alpha)$$

which is non-negative if $\mathcal{A} = 0$ or $\mathcal{B} = 0$. If $\mathcal{A}, \mathcal{B} > 0$ then $i'(\bar{p})$ and $j'(\bar{q}/\rho)$ are finite and then from (5.10) and (5.11) we obtain

$$1 - (2\mathcal{M} - \mathcal{N}) = \frac{(1 - \alpha)(1 - \beta)(i'(\bar{p})j'(\bar{q}/\rho) - 1)}{(1 - (\bar{p}/\bar{q})i'(\bar{p}))(1 - (\bar{q}/\bar{p})j'(\bar{q}/\rho))} \geqslant 0,$$

$$(5.12)$$

on account of (5.8). Thus in all cases $\mathcal{M} \leqslant (\mathcal{N} + 1)/2$ and it follows that $\lambda_1 \leqslant 1$. This completes the proof of (i).

(ii) Under the conditions of this part of the theorem, $\mathcal{A}, \mathcal{B} > 0$ and hence $\mathcal{D} > 0$. It follows that $\lambda_2 < \lambda_1$. In addition (5.12) implies that $\mathcal{M} < (1 - \mathcal{N})/2$ and hence $\lambda_1 < 1$.

Letting

$$D = \alpha + \mathcal{A}, \qquad C = (1 - \alpha)\bar{z}^2 F'(\bar{z})/\bar{p},$$

$$\Delta = \beta + \mathcal{B} \quad \text{and} \quad \Gamma = (1 - \beta)(1 - \bar{z})^2 \rho G'(1 - \bar{z})/\bar{q},$$

it follows from (5.10) and (5.11) that $D, \Delta < 1$ and

$$\mathcal{J}(\varepsilon) = \begin{bmatrix} D & -C \\ -\Gamma & \Delta \end{bmatrix}.$$

It is easy to check that $\lambda_1 > D \vee \Delta$ and $\lambda_2 < D \wedge \Delta$ and the asserted qualitative structure of the eigenvalues follows.

Now C can be expressed in terms of $i'(\bar{p})$ and this expression can be combined with (5.10) to yield $-i'(\bar{p}) = (1 - D)/C$. This shows that $C > 0$. Solving the equation $\mathcal{J}e_1 = \lambda_1 e_1$, where e_1 has the form given in the theorem, yields $u = (\lambda_1 - D)/C$ and hence $u + i'(\bar{p}) = (\lambda_1 - 1)/C < 0$. The right-hand side of (5.9) follows. Similarly it can be shown that $-j'(\bar{q}/\rho) = (1 - \Delta)/\Gamma$ and $1/u = (\lambda_1 - \Delta)/\Gamma$, giving the left-hand member of (5.9). The proof is complete.

Corollary 5.4.

Under the conditions of Theorem 5.3 (ii)

$$(1 - D)(1 - \Delta) - C\Gamma > 0. \qquad (5.13)$$

Proof. The various definitions above give $2\mathcal{M} = D - \Delta$ and $\mathcal{N} = D\Delta - C\Gamma$. Inequality (5.13) is just another form of $2\mathcal{M} < 1 + \mathcal{N}$.

Under the conditions of Theorem 5.3 (ii) the linearization theorem (see §3.2) suggests that the dynamical system $\{(p_n - \bar{p}, q_n - \bar{q})\}$ behaves asymptotically as does the linearized system $\{\delta_n\}$ where $\delta_{n+1} = \mathcal{J}\delta_n$ and $\mathcal{J} = \mathcal{J}(\varepsilon)$. Standard theory shows that provided $\delta_1 \neq \kappa\mathbf{e}_2$ for any κ then δ_n tends to $\mathbf{0}$ along $\mathbf{e}_1$ at a geometric rate. Our next theorem proves that these features are essentially true of our nonlinear system. The proof comprises two parts, the first being to show that both $\{\lambda_1^{-n}(p_n - \bar{p})\}$ and $\{\lambda_1^{-n}(q_n - \bar{q})\}$ converge, and the second to determining the sign of the limits. Let $\varepsilon_n = p_n - \bar{p}$ and $\eta_n = q_n - \bar{q}$.

> *Theorem 5.5.*
>
> *In addition to the assumptions of Theorem 5.3 (ii), suppose that $F''(z)$ and $G''(z)$ exist and are bounded in an open interval containing $\bar{z}$. There is a constant μ such that $\lambda_1^{-n}\varepsilon_n \to \mu$ and $\lambda_1^{-n}\eta_n \to -\mu u$. In addition there is a positive constant γ such that for all sufficiently large n*

$$\lambda_1^{-n}\left(1 - \frac{\gamma\lambda_1^n}{1 - \lambda_1}\right)\frac{\varepsilon_n - v\eta_n}{1 + uv} \leqslant \mu \leqslant \lambda_1^{-n}\left(\exp\frac{2\gamma\lambda_1^n}{1 - \lambda_1}\right)\frac{\varepsilon_n - v\eta_n}{1 + uv}$$

(5.14)

if $\varepsilon_n - v\eta_n > 0$, and the reverse inequality holds if $\varepsilon_n - v\eta_n < 0$.

Remarks. When conditions (5.3) are satisfied for Model 1A then the interior equilibrium, given by (5.4), is GAS and all the conditions required for Theorem 5.5 are satisfied. From (4.7) we see that $i(p) = c(1 - p)$ and $j(q) = d(\rho - q)$. Using these and (3.4), a little algebra will show that the elements of $\mathcal{J}(\varepsilon)$ are given via

$$D = \alpha + (1 - \alpha)(\bar{q}/c), \quad C = (1 - \alpha)(\bar{p}/c)$$
$$\Gamma = (1 - \beta)(\bar{q}/\rho d), \quad \Delta = \beta + (1 - \beta)(\bar{p}/\rho d).$$

The eigenvalue λ_1 determining the rate of approach to ε does not reduce to a simple form – it is given by $\lambda_1 = \mathcal{M} + \mathcal{P}$ where

$$\mathcal{M} = [\alpha + \beta + (1 - \alpha)\bar{q}/c + (1 - \beta)(\bar{p}/\rho d)]/2$$

and

$$\mathcal{P} = (\mathcal{M}^2 - \mathcal{N})^{1/2}$$
$$= \tfrac{1}{2}\{[\alpha - \beta + (1 - \alpha)(\bar{q}/c) - (1 - \beta)(\bar{p}/\rho d)]^2$$
$$+ (1 - \alpha)(1 - \beta)(\bar{p}\bar{q}/\rho cd)\}^{1/2}.$$

We have seen that $\mathcal{M} \leqslant (\mathcal{N} + 1)/2$ and this gives the upper bound

$$\lambda_1 < 1 - \frac{(1-\alpha)(1-\beta)}{2}\left(\frac{(\rho d - 1)(c - \rho)}{\rho(cd - 1)}\right).$$

Several lower bounds can be obtained from the bound $\mathcal{P}^2 > [|\alpha - \beta| - (D - \alpha) - (\Delta - \beta)]^2/4$. Thus, if $\alpha > \beta$ then $\lambda_1 > \alpha$ when $2\alpha > D + \Delta$ and $\lambda_1 > D + \Delta - \alpha > 0$ when $2\alpha < D + \Delta$.

In the discussion at the end of this section we show that, in general, orbits approach ε along the direction determined by the eigenvalue $\mathbf{e}_1 = (1, -u)^T$. For Model 1A we have

$$u = [c(\lambda_1 - \alpha)/(1 - \alpha) - \bar{q}]/\bar{p},$$

and the eigenvalue $\mathbf{e}_2 = (v, 1)^T$ is determined by

$$v = \frac{1-\alpha}{1-\beta}(\rho\,d\bar{p}u/c\bar{q}).$$

Suppose that ρ, c and d are held constant and α and β are allowed to vary. If $\alpha = 1$ then $u = 1/d$ and $v = 0$; if $\alpha = \beta$ then $u = c/\rho d$ and $v = \bar{p}/\bar{q}$; if $\beta = 1$ then $u = c$ and $v = \infty$. It follows from the intermediate value theorem that $\mathbf{e}_1$ can point in any south-easterly direction between the isoclines. Suppose α is small but β is chosen to be close to unity, then fixed. In this case $\mathbf{e}_1$ points in approximately the same direction as $\mathcal{I}_F$ and $\mathbf{e}_2$ points very slightly north of due east, that is, it makes a small positive angle with the Op direction. As α increases, $\mathbf{e}_1$ rotates toward $\mathcal{I}_G$ and $\mathbf{e}_2$ rotates toward due north. When $\alpha = \beta$, $\mathbf{e}_2$ lies along the ray through ε. As α approaches 1, the direction of $\mathbf{e}_1$ approaches that of $\mathcal{I}_G$ and the direction of $\mathbf{e}_2$ approaches due north.

Proof. We begin the proof of the convergence assertion by showing that $\Sigma |\varepsilon_n|$ and $\Sigma |\eta_n|$ are finite. Since $\bar{p} = F(\bar{z})$, the mean value theorem yields the recurrence

$$\varepsilon_{n+1} = D_n\varepsilon_n - C_n\eta_n \tag{5.15}$$

where

$$D_n = \alpha + (1 - \alpha)(1 - \bar{z})z_n F'(z_n')/p_n, \quad C_n = (1 - \alpha)\bar{z}z_n F'(z_n')/p_n$$

and z_n' is between $\bar{z}$ and z_n. Similarly

$$\eta_{n+1} = -\Gamma_n\varepsilon_n + \Delta_n\eta_n \tag{5.16}$$

where

$$\Delta_n = \beta + (1 - \beta)(1 - z_n)\bar{z}\rho G'(1 - z_n'')/q_n,$$

$$\Gamma_n = (1 - \beta)(1 - z_n)\rho G'(1 - z_n'')/q_n$$

and z_n'' is between $\bar{z}$ and z_n. The coefficients in (5.15) and (5.16) converge to the elements of $\mathscr{J}$ and from (5.13) it follows that there exists $\delta > 0$ and n' such that $V_\delta = (1 - D - \delta)(1 - \Delta - \delta) - (C + \delta)(\Gamma + \delta) > 0$, $\delta + \Delta < 1$, $D_n \leqslant D + \delta$, $\Delta_n \leqslant \Delta + \delta$, $C_n \leqslant C + \delta$ and $\Gamma_n \leqslant \Gamma + \delta$ whenever $n \geqslant n'$. It follows from (5.15) that for such n we have $|\varepsilon_{n+1}| \leqslant (D + \delta)|\varepsilon_n| + (C + \delta)|\eta_n|$, whence

$$(1 - D - \delta) \sum_{j=n'}^{n} |\varepsilon_j| \leqslant (C + \delta) \sum_{j=n'}^{n} |\eta_j| + |\varepsilon_{n'}| - |\varepsilon_{n+1}|.$$

Combining this with the analogous inequality obtained from (5.16) yields

$$V_\delta \sum_{j=n'}^{n} |\varepsilon_j| \leqslant (1 - \Delta - \delta)(|\varepsilon_{n'}| - |\varepsilon_{n+1}|) + |\eta_{n'}| - |\eta_{n+1}|.$$

Allowing $n \to \infty$ yields $\Sigma_{j=n'}^{\infty} |\varepsilon_j| < \infty$ and the inequality obtained from (5.16) then implies that $\Sigma_{j=n'}^{\infty} |\eta_j| < \infty$.

We now show that for any $\varepsilon > 0$, $(\lambda_1 + \varepsilon)^{-n} \varepsilon_n \to 0$ and $(\lambda_1 + \varepsilon)^{-n} \eta_n \to 0$. The canonical form of $\mathscr{J}$ is given by $\Lambda = \text{diag}(\lambda_i) = \mathscr{P}^{-1} \mathscr{J} \mathscr{P}$ where $\mathscr{P} = [\mathbf{e}_1 : \mathbf{e}_2]$. By letting

$$\Lambda_n = \mathscr{P}^{-1} \begin{bmatrix} D_n & -C_n \\ -\Gamma_n & \Delta_n \end{bmatrix} \mathscr{P} \quad \text{and} \quad \zeta_n = (\tilde{\varepsilon}_n, \tilde{\eta}_n)^T = \mathscr{P}^{-1}(\varepsilon_n, \eta_n)^T$$

we can express (5.15) and (5.16) as

$$\zeta_{n+1} = \Lambda_n \zeta_n \quad (n \geqslant 1). \tag{5.17}$$

Now $\Lambda_n \to \Lambda$ and hence for each $\varepsilon > 0$ such that $\lambda_1 + \varepsilon < 1$ there exists n'' for which we have

$$|\zeta_{n+1}| \leqslant \Lambda_\varepsilon |\zeta_n| \quad (n \geqslant n'') \tag{5.18}$$

where $|\zeta_n| = (|\tilde{\varepsilon}_n|, |\tilde{\eta}_n|)^T$ and

$$\Lambda_\varepsilon = \begin{bmatrix} \lambda_1 + \varepsilon/2 & \varepsilon/2 \\ \varepsilon/2 & \lambda_2 + \varepsilon/2 \end{bmatrix}.$$

Iteration of (5.18) yields $|\zeta_{n+n''}| \leqslant \Lambda_\varepsilon^n |\zeta_{n''}|$. But Λ_ε is a positive matrix with maximal eigenvalue λ_ε satisfying $\lambda_1 + \varepsilon/2 < \lambda_\varepsilon < \lambda_1 + \varepsilon$. The Perron–Frobenius theorem (Gantmacher (1959)) shows that $\Lambda_\varepsilon^n = O(\lambda_\varepsilon^n)$, whence $(\lambda_1 + \varepsilon)^{-n} |\zeta_n| \to 0$.

Now let $\varepsilon_n^* = \lambda_1^{-n} \tilde{\varepsilon}_n$, $\eta_n^* = \lambda_1^{-n} \tilde{\eta}_n$ and $\theta = \lambda_2 / \lambda_1$. We can write (5.17) as

$$\varepsilon_{n+1}^* = \varepsilon_n^* + r_n', \quad \eta_{n+1}^* = \theta \eta_n^* + r_n'' \quad (n \geqslant 1)$$

where

$$(r_n', r_n'')^T = \lambda_1^{-n-1}(\Lambda_n - \Lambda)\zeta_n.$$

Using the second derivative assumptions we see from its construction that $\Lambda_n = \Lambda + O(|\varepsilon_n| + |\eta_n|)$ whence r'_n and r''_n are $O[(\lambda_1^{-1}(\lambda_1 + \varepsilon)^2)^n]$. By choosing ε small enough we can ensure that $\lambda_1^{-1}(\lambda_1 + \varepsilon)^2 < 1$ and it follows that both $\Sigma r'_n$ and $\Sigma r''_n$ converge absolutely. We conclude that

$$\mu = \lim_{n \to \infty} \varepsilon_n^* = \varepsilon_1^* + \sum_{n=1}^{\infty} r'_n < \infty.$$

It follows that $r''_n = O(\lambda_1^n)$ whence

$$\eta_n^* = \eta_1^* \theta^{n-1} + \sum_{j=1}^{n-1} r''_j \theta^{n-1-j} = \begin{cases} O(\lambda_1^n) & \text{if } \lambda_1^2 > \lambda_2 \\ O(n\lambda_1^n) & \text{if } \lambda_1^2 = \lambda_2 \\ O(\theta^n) & \text{if } \lambda_1^2 < \lambda_2 \end{cases}.$$

In all cases $\eta_n^* \to 0$ whence $\mathscr{P}^{-1}(\varepsilon_n, \eta_n)^T \sim \lambda_1^n(\mu, 0)^T$ and the convergence assertion follows.

We now turn to the task of obtaining estimates of μ. Iteration of (5.17) yields

$$\zeta_{n+v} = (1 + uv)^{-1} \prod_{j=n}^{n+v-1} \Lambda_j(\varepsilon_n - v\eta_n, u\varepsilon_n + \eta_n)^T. \tag{5.19}$$

If n is large enough there is a constant $\gamma > 0$ such that $1 - \gamma\lambda_1^n > 0$ and for $j \geqslant n$,

$$\tilde{\Lambda}_j = \lambda_1^{-1}\Lambda_j \geqslant \begin{bmatrix} 1 - \gamma\lambda_1^j & -\gamma\lambda_1^j \\ -\gamma\lambda_1^j & \theta - \gamma\lambda_1^j \end{bmatrix}$$

and it can be shown by induction that

$$\prod_{j=n}^{n+v-1} \tilde{\Lambda}_j \geqslant \begin{bmatrix} 1 - \gamma\lambda_1^n(1 - \lambda_1^v)/(1 - \lambda_1) & -\gamma\lambda_1^n(\theta^v - \lambda_1^v)/(\theta - \lambda_1) \\ -\gamma\lambda_1^n(1 - (\theta\lambda_1)^v)/(1 - \theta\lambda_1) & \theta^v - \gamma\lambda_1^n(1 - \lambda_1^v)/(1 - \lambda_1) \end{bmatrix}.$$

$$\tag{5.20}$$

An upper bound for the product can be obtained as follows. For sufficiently large n there is a positive constant which can be taken as γ such that for $j \geqslant n$

$$\tilde{\Lambda}_j \leqslant \mathscr{U}_j = \begin{bmatrix} 1 + \gamma\lambda_1^j & \gamma\lambda_1^j \\ \gamma\lambda_1^j & \theta + \gamma\lambda_1^j \end{bmatrix}.$$

The Lagrange–Sylvester interpolation theorem applied to a 2×2 matrix M with distinct eigenvalues ϕ_1, ϕ_2 and a function f yields

$$f(M) = (\phi_1 - \phi_2)^{-1}[f(\phi_1)(M - \phi_2\mathscr{I}) - f(\phi_2)(M - \phi_1\mathscr{I})]$$

where $\mathcal{I}$ is the identity matrix; see Gantmacher (1959, p. 97). We shall use this formula with $f(x) = \log x$ and then $f(x) = e^x$ to obtain an upper bound for the product

$$\prod_{j=n}^{n+v-1} \mathcal{U}_j = \exp \sum_{j=n}^{n-v-1} \log \mathcal{U}_j$$

Writing $a = \gamma \lambda_1^j$, the eigenvalues $v' > v''$ of $\mathcal{U}_j$ satisfy $1 + a < v' < 1 + 2a$ and $\theta < v'' < \theta + a$. It follows that $v' - v'' > 1 - \theta$, $\log v' < 2a$ and $\log v'' < \log \theta$ whence

$$\log \mathcal{U}_j \leqslant \begin{bmatrix} 2\gamma \lambda_1^j & \tau \gamma \lambda_1^j + \dfrac{2\gamma^2 \lambda_1^{2j}}{1-\theta} \\[2ex] \tau \gamma \lambda_1^j + \dfrac{2\gamma^2 \lambda_1^{2j}}{1-\theta} & \log \theta + \dfrac{2\gamma^2 \lambda_1^{2j}}{1-\theta} \end{bmatrix}$$

where $\tau = -(1-\theta)^{-1} \log \theta$. Thus

$$\sum_{j=n}^{n+v-1} \log \mathcal{U}_j \leqslant$$

$$\leqslant \begin{bmatrix} 2\gamma \lambda_1^n \dfrac{1-\lambda_1^v}{1-\lambda_1} & \tau \gamma \lambda_1^n \cdot \dfrac{1-\lambda_1^v}{1-\lambda_1} + \dfrac{2\gamma^2 \lambda_1^{2n}}{1-\theta} \cdot \dfrac{1-\lambda_1^{2v}}{1-\lambda_1^2} \\[2ex] \tau \gamma \lambda_1^n \dfrac{1-\lambda_1^v}{1-\lambda_1} + \dfrac{2\gamma^2 \lambda_1^{2n}}{1-\theta} \cdot \dfrac{1-\lambda_1^{2v}}{1-\lambda_1^2} & v \log \theta + \dfrac{2\gamma^2 \lambda_1^{2n}}{1-\theta} \cdot \dfrac{1-\lambda_1^{2v}}{1-\lambda_1^2} \end{bmatrix}.$$

The right-hand side has the form

$$\begin{bmatrix} \varepsilon & \varepsilon V \\ \varepsilon V & -U + \varepsilon W \end{bmatrix}$$

where

$$\varepsilon = 2\gamma \lambda_1^n \frac{1-\lambda_1^v}{1-\lambda_1}$$

and $U = v \log \theta^{-1}$ is large when v is large. This matrix has distinct eigenvalues ℓ', ℓ'' satisfying $\varepsilon < \ell' < \varepsilon(1 + V)$, $-U + \varepsilon(W - V) < \ell'' < -U + \varepsilon W$ and as $U \to \infty$, $\ell' - \ell'' \to \infty$ and $(\varepsilon - \ell'')/(\ell' - \ell'') \to 1$. The Lagrange–Sylvester formula yields

$$\exp \begin{bmatrix} \varepsilon & \varepsilon V \\ \varepsilon V & -U + \varepsilon W \end{bmatrix} = \frac{\exp \ell'}{\ell' - \ell''} \begin{bmatrix} \varepsilon - \ell'' & \varepsilon V \\ \varepsilon V & -U + \varepsilon W - \ell'' \end{bmatrix}$$

$$+ \frac{\exp \ell''}{\ell' - \ell''} \begin{bmatrix} \varepsilon - \ell' & \varepsilon V \\ \varepsilon V & -U + \varepsilon W - \ell' \end{bmatrix} \sim e^{\ell'} \begin{bmatrix} 1 & 0 \\ 0 & 0 \end{bmatrix}$$

as $U \to \infty$, since ε, V and W remain bounded as $v \to \infty$. Calculation of the explicit form of ℓ' shows that $\ell' \to 2\gamma\lambda_1^n/(1 - \gamma_1)$ as $v \to \infty$ and hence

$$\Phi(n) = \prod_{j=n}^{\infty} \tilde{\Lambda}_j \leqslant \left(\exp\frac{2\gamma\lambda_1^n}{1 - \lambda_1} \right)\begin{bmatrix} 1 & 0 \\ 0 & 0 \end{bmatrix}.$$

Allowing $v \to \infty$ in (5.20), the upper bound just obtained shows that $\Phi_{12}(n) = 0$ and (5.19) yields

$$\mu = \frac{\lambda_1^{-n}}{1 + uv}\Phi_{11}(n)(\varepsilon_n - v\eta_n).$$

Inequality (5.14) and its reversed form follow from the bounds for $\Phi_{11}(n)$ established above. This completes the proof.

We can now determine how orbits approach ε. The bounds established above show that $\Phi_{11}(n) > 0$ for all sufficiently large n. It follows that $\mu = 0$ iff $q_n = \bar{q} + (p_n - \bar{p})/v$ and this could correspond to at most one orbit. All other orbits must have $\mu \neq 0$ and the convergence assertion of Theorem 5.5 shows that they approach ε asymptotically toward a line through ε in the direction of $\mathbf{e}_1$. In particular all orbits, with at most one exception, approach ε from within $\mathcal{R}_2$ or $\mathcal{R}_4$.

5.5 Unstable interior equilibrium

In this section we consider the situation typified by Figure 5.3. More specifically, recalling that for each $z \in (0, 1)$, R_z meets $\mathcal{J}_F$ and $\mathcal{J}_G$ exactly once, we assume when $z < \bar{z}$ (respectively, $z > \bar{z}$) that $R_z \cap \mathcal{J}_F$ is closer to (respectively, further from) $(0, 0)$ than $R_z \cap \mathcal{J}_G$. It should be clear that these conditions are equivalent to

$$F(z) < (>)\frac{z}{1 - z}\rho G(1 - z) \quad \text{if} \quad 0 < z < \bar{z} \text{ (respectively, } \bar{z} < z < 1).$$

$$(5.21)$$

The following theorem allows us to deduce the behavior of the orbits under these conditions.

Theorem 5.6.

Suppose there exists exactly one equilibrium point $\varepsilon \in \mathcal{S}^0$ and that (5.21) holds. Then $\mathcal{R}_2$, $\mathcal{R}_4$, $\mathcal{R}_1 \cup \mathcal{R}_2 \cup \mathcal{R}_4$ and $\mathcal{R}_3 \cup \mathcal{R}_2 \cup \mathcal{R}_4$ are positively invariant. In addition

$$i'(p)j'(\bar{q}/\rho) \leqslant 1 \tag{5.22}$$

and ε is unstable if strict inequality holds at (5.22).

Proof. The invariance assertions are proved as for Theorem 5.1. To see (5.22) consider the following cases.

If $j'(\bar{q}/\rho) \geqslant 0$ then $i'(\bar{p}) \leqslant 0$ and (5.22) must be valid.

If $j'(\bar{q}/\rho) < 0$ then (5.22) is certainly satisfied if $i'(\bar{p}) \geqslant 0$.

On the other hand if $i'(\bar{p}) < 0$ then in a sufficiently small open neighborhood of $\bar{p}$, $\mathscr{I}_F$ crosses $\mathscr{I}_G$ from below as p increases through $\bar{p}$. The slope of the normal vector to $\mathscr{I}_G$ at ε is, when referred to the (p, q)-co-ordinate system, equal to $-j'(\bar{q}/\rho)$ and this cannot exceed the corresponding quantity for $\mathscr{I}_F$, namely, $-i'(\bar{p})$. This yields (5.22).

Some simple sketches will convince the reader that $i'(\bar{p})$ and $j'(\bar{q}/\rho)$ can take either sign (specifically, $-\infty \leqslant i'(\bar{p}) \leqslant \bar{q}/\bar{p}$ and $-\infty \leqslant j'(\bar{q}/\rho) \leqslant \bar{p}/\bar{q}$) and hence the inequalities (5.10) and (5.11)' must be replaced by $0 \leqslant \mathscr{A}, \mathscr{B} \leqslant \infty$. It follows that $\alpha + \beta \leqslant 2\mathscr{M} \leqslant \infty$ and hence that the larger eigenvalue $\lambda_1 = \mathscr{M} + (\mathscr{M}^2 - \mathscr{N})^{1/2} > 1$ if $\mathscr{M} \geqslant 1$. Suppose $\mathscr{M} < 1$. Then

$$\lambda_1 - 1 = \sqrt{\mathscr{M}^2 - \mathscr{N}} - (1 - \mathscr{M}) = \frac{2\mathscr{M} - \mathscr{N} - 1}{\sqrt{\mathscr{M}^2 - \mathscr{N}} + 1 - \mathscr{M}},$$

and it follows from (5.12) that $\lambda_1 \geqslant 1$ iff (5.22) holds. But we have just seen that it always holds. When strict inequality holds at (5.22) we have $\lambda_1 > 1$ and then ε is unstable.

Theorem 5.6 allows us to make some informed conjectures about orbit configurations. We shall restrict our discussion to the case where the isoclines intersect without osculating, in which case inequality (5.22) is strict, $\lambda_1 > 1$, and ε is unstable. Obviously Model 1A satisfies this condition. In addition $\lambda_2 < 1$ and hence, following the terminology of Guckenheimer and Holmes (1983, p. 17), ε is a hyperbolic fixed point. According to the stable manifold theorem (op. cit. p. 18) there exists a set of points $\mathscr{M}_s \in \mathscr{S}$, the local stable manifold, which is tangent to the eigenvector of λ_2 at ε, contains ε and if $\mathbf{x} \in \mathscr{M}_s$ then $\mathscr{T}^n(\mathbf{x}) \to \varepsilon$ ($n \to \infty$). Orbits starting from any point not in $\mathscr{M}_s$ do not converge to ε. Unfortunately, we cannot see how to determine $\mathscr{M}_s$ except when $\alpha = \beta$, and then $\mathscr{M}_s \in R_{\bar{z}}$.

It is a reasonable conjecture that $\mathscr{M}_s$ is the common boundary of two *domains of attraction*. $\mathscr{D}_p$ and $\mathscr{D}_q$ such that, for example, if $\mathbf{x} \in \mathscr{D}_p$ than $\mathscr{T}^n(\mathbf{x}) \to (1, 0)$. Certainly if $\mathbf{x} \in \mathscr{R}_2$ then $\mathscr{T}^n(\mathbf{x}) \to (1, 0)$ and it seems likely that orbits starting in $\mathscr{R}_1 \backslash \mathscr{M}_s$ eventually hop into either $\mathscr{R}_2$ or $\mathscr{R}_4$ and then converge to one of the boundary equilibria. Similarly orbits starting in $\mathscr{R}_3 \backslash \mathscr{M}_s$ should either enter $\mathscr{R}_2$ or $\mathscr{R}_4$, or converge directly to one of the boundary equilibria.

For Model 1A, Theorem 4.2 can be applied in small regions containing the boundary equilibria to decide whether converging orbits must enter $\mathcal{R}_2 \cup \mathcal{R}_4$, or not. The proof of Theorem 4.2 is in essence a local analysis carried out near $(0, \rho)$ and its key ingredient is simply the assumption that near $(0, \rho)$, $\mathcal{I}_G$ lies above $\mathcal{I}_F$. An analogous discussion can be developed for orbits converging to $(1, 0)$. As a consequence we can infer that orbits converging to the boundary equilibria do so from within $\mathcal{R}_2$ or $\mathcal{R}_4$ iff

$$\beta < (1 - \alpha)c/\rho \quad and \quad \alpha < (1 - \beta)\rho d + \beta. \tag{5.23}$$

Still considering Model 1A, it is clear that an unstable internal equilibrium exists iff

$$\rho > c \quad and \quad \rho d < 1 \quad (\text{whence } cd < 1). \tag{5.24}$$

Let us assume $\alpha > \beta$ and determine the configuration of $\mathcal{C} = \{\mathbf{x} : \sigma(p, q) = 0\}$; see (4.4). As in Chapter 4 this is most easily approached by transforming to (x, y)-coordinates in $\mathcal{W}$. Since $c < 1/d$ the image of Op now lies above that of Oq (compare Figure 5.5). The image of $\mathcal{C}$ in $\mathcal{W}$ is still the rectangular hyperbola defined by (4.11) and its relation to the boundaries of $\mathcal{W}$ and the isoclines is as shown in Figure 5.5 except that the boundary rays interchange, that is, the line $y = x/d$ now lies above the line $y = cx$.

We immediately infer that $\mathcal{C}$ is a curve linking $(p(1), 0)$, ε and $(0, q(0))$ where

$$p(1) = 0 \quad \text{if} \quad \frac{1 - \beta}{1 - \alpha}\rho d \leqslant 1$$

and in the contrary case $p(1)$ is given by (4.14a), and $q(0)$ is given by (4.14b). When $p(1) > 0$ then $\mathcal{C}$ meets Op perpendicularly when $\psi = \rho cd^2(1 - \beta)/(1 - \alpha) = 1$. As $\alpha \to 1$, $\mathcal{C}$ tends to $\mathcal{I}_G$. Some typical configurations are shown in Figure 5.8.

Some typical orbits are illustrated in Figure 5.9 for the case where (5.23) holds. Note that we have not actually proved that the behavior illustrated is typical, but our analysis, and numerical examples we have examined, strongly suggest that this is so.

Estimates of $\mathcal{D}_p$ and $\mathcal{D}_q$ for Model 1A can be obtained using a Liapunov function as follows – see Goh ((1980), Theorem 3.14.1). Consider the equilibrium point $(0, \rho)$ and the function

$$V(\mathbf{x}) = \frac{p}{1 - \alpha} + d\frac{q - \rho - \rho \log(q/\rho)}{1 - \beta}.$$

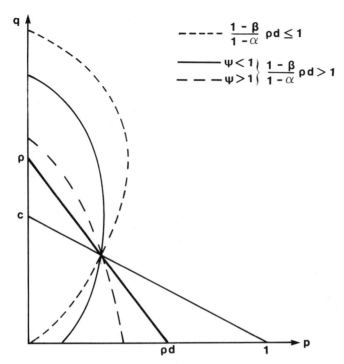

Figure 5.8. Model 1A: Possible configurations of $\mathscr{C}$ when $\alpha > \beta$.

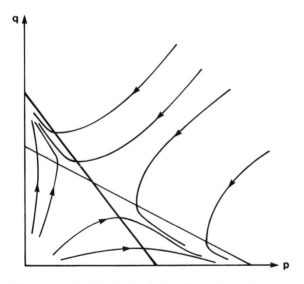

Figure 5.9. Model 1A: Typical orbit configurations for an unstable internal equilibrium.

Clearly $V(0, \rho) = 0$ but $V(\mathbf{x}) > 0$ for all other $\mathbf{x} \in \mathbb{R}_+^2$. Now

$$\dot{V} = p\left(\frac{c}{cp + q} - 1\right) + dq\left(\frac{\rho d}{p + dq} - 1\right)$$

$$- \frac{\rho d}{1 - \beta} \log\left(\beta + (1 - \beta)\frac{\rho d}{p + dq}\right)$$

$$\leqslant \frac{cp}{cp + q} + \frac{pd^2 q}{p + dq} - pd = p\left(\frac{c}{cp + q} - \frac{\rho d}{p + qd}\right),$$

and we have used (5.7) to obtain the bound. It follows that $\dot{V} \leqslant 0$ if $q \geqslant p\bar{q}/\bar{p}$ and the first equality is strict unless $\mathbf{x} = (0, \rho)$ or ε.

Let $u > 0$ and $\mathscr{V}(u) = \{\mathbf{x} : V(\mathbf{x}) < u\}$. The set $\mathscr{V}(u)$ is positively invariant and if $\dot{V}(\mathbf{x}) < 0$ whenever $\mathbf{x} \in \mathscr{V}(u) \backslash \{(0, \rho)\}$ then $\mathscr{V}(u) \subset \mathscr{D}_q$. Thus we should choose u to be the largest number for which $\mathscr{V}(u) \subset \{\mathbf{x} : q \geqslant p\bar{q}/\bar{p}\}$. Letting $\lambda(q) = q - \rho - \rho\log(q/p)$ it is obvious that

$$\mathscr{V}(u) = \left\{\mathbf{x} : \frac{p}{1 - \alpha} \leqslant u - \frac{d\lambda(q)}{1 - \beta}\right\}$$

and hence $\mathscr{V}(u_1) \subset \mathscr{V}(u_2)$ if $u_1 < u_2$. It follows that the largest permissible $\mathscr{V}(u)$ is obtained by adjusting u so that the curved boundary of $\mathscr{V}(u)$ osculates $R_{\tilde{z}}$.

Let $0 < q_1 < \rho < q_2 < \infty$ be solutions of the equation $\lambda(q) = u(1 - \beta)/d$. The curved boundary of $\mathscr{V}(u)$ is $\{\mathbf{x} : p = (1 - \alpha)(u - d\lambda(q)/(1 - \beta)), q_1 \leqslant q \leqslant q_2\}$ and hence the osculation condition is met if $\tilde{q}$ is chosen so that $-[(1 - \alpha)d/(1 - \beta)]\lambda'(\tilde{q}) = \bar{p}/\bar{q}$, that is,

$$\tilde{q} = \rho \bigg/ \left[1 + \frac{(1 - \beta)\bar{p}}{(1 - \alpha)d\tilde{q}}\right] < \rho$$

and then choosing

$$u = \frac{d\lambda(\tilde{q})}{1 - \beta} + \frac{\tilde{q}\bar{p}}{(1 - \alpha)\tilde{q}}.$$

Observe that $\tilde{q} < \bar{q}$ iff $\alpha > \beta$. The region of attraction estimated in this way is shown in Figure 5.10.

Similarly, by starting with

$$U(\mathbf{x}) = c\frac{p - 1 - \log p}{1 - \alpha} + \frac{q}{1 + \beta}$$

and

$$\mathscr{U}(a) = \{\mathbf{x} : q \leqslant (1 - \beta)/a - cv(p)/(1 - \alpha)\}$$

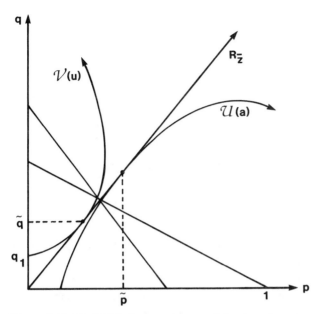

Figure 5.10. Model 1A: Inner estimates of the domains of attraction.

where $v(p) = p - 1 - \log p$, we find that $\mathcal{U}(a) \subset \mathcal{D}_p$ if $a = cv(\tilde{p})/$
$(1 - \alpha) + \tilde{p}\bar{q}/(1 - \beta)\bar{p}$ where

$$\tilde{p} = [1 + (1 - \alpha)\bar{q}/(1 - \beta)c\bar{p}]^{-1}$$

and $\tilde{p} > \bar{p}$ iff $\alpha > \beta$.

5.6   Coincident isoclines

For Model 1A the only remaining case is that where
$c = \rho = d^{-1}$ and then $\mathcal{I}_F = \mathcal{I}_G$. Even for Model G we can obtain a
substantially complete qualitative understanding when the isoclines
coincide. In this case (3.16) and (3.17) are functional specifications of
the one curve and it follows that the function $i(\cdot)$ defined by (3.16) is
non-increasing. Now $\mathcal{I}_F$ is specified by (3.15) and (3.16) and together
these show that the identity

$$G(z) = \frac{zF(1 - z)}{\rho(1 - z)} \quad (0 \leqslant z \leqslant 1) \tag{5.25}$$

is necessary and sufficient for $\mathcal{I}_F$ and $\mathcal{I}_G$ to coincide.

Equations (3.7) can be written as

$$p_{n+1} = (1 - \alpha)F(z_n) + \alpha p_n, \quad q_{n+1} = (1 - \beta)(q_n/p_n)F(z_n) + \beta q_n$$

$$\tag{5.26}$$

which imply that

$$\frac{q_{n+1} - q_n}{p_{n+1} - p_n} = \frac{1 - \beta}{1 - \alpha} \cdot \frac{q_n}{p_n}. \tag{5.27}$$

When $\alpha = \beta$ this yields $q_{n+1}/p_{n+1} = q_n/p_n$, whence $q_n = p_n(q_1/p_1)$. Thus the orbit starting at (p_1, q_1) is contained in R_{z_1}, where $z_1 = p_1/(p_1 + q_1)$. The first of (5.26) becomes $p_{n+1} = (1 - \alpha)F(z_1) + \alpha p_n$ from which we obtain

$$p_n = p_1 \alpha^{n-1} + (1 - \alpha^{n-1})F(z_1) \quad (n \geqslant 1).$$

In particular

$$p_n \to F(z_1) \quad \text{and} \quad q_n \to (q_1/p_1)F(z_1),$$

the rate of convergence is proportional to α^n, and the orbit approaches $\mathscr{I}_F$ from above iff **x** lies above $\mathscr{I}_F$.

When $\alpha \neq \beta$ the qualitative picture above persists in the sense made clear in the following theorem. We continue to use the notation for the regions $\mathscr{R}_1$, and so on, except now $\mathscr{R}_2 = \mathscr{R}_4 = \phi$.

> *Theorem 5.7.*
> *Suppose (5.25) is satisfied.*
> (i) *The regions $\mathscr{R}_1$ and $\mathscr{R}_3$ are positively invariant and the curve $\mathscr{C}$ coincides with $\mathscr{I}_F$. In addition, $\mathscr{R}_p = \mathscr{R}_3$ and $\mathscr{R}_q = \mathscr{R}_1$ (respectively, $\mathscr{R}_p = \mathscr{R}_1$ and $\mathscr{R}_q = \mathscr{R}_3$) if $\alpha > \beta$ (respectively, if $\alpha < \beta$).*
> (ii) *If $(p_1, q_1) \in \mathscr{S}^0$ then $(p_n, q_n) \to (p, q) \in \mathscr{I}_F$ and the convergence is monotonic.*
> (iii) *Each point of $\mathscr{I}_F$ is neutrally stable.*
> (iv) *If $(p, q) \in \mathscr{I}_F$ and $0 < p < 1$, then*
>
> $$\gamma = \alpha + (\beta - \alpha)qF'(z)/(p + q)^2 \leqslant \alpha \vee \beta \tag{5.28}$$
>
> *and equality holds iff $\alpha = \beta$. Suppose $(p_n, q_n) \to (p, q)$, $F'(z) > 0$ and $|F''(z)| < \infty$. There exists a finite number L, depending on (p_1, q_1), such that as $n \to \infty$,*
>
> $$p_n - p \sim L\gamma^n \quad \text{and} \quad q_n - q \sim \frac{1 - \beta}{1 - \alpha} \cdot \frac{q}{p} L\gamma^n.$$

Remarks. Each orbit converges to some point on $\mathscr{I}_F$ which depends on the initial point of the orbit. If this point is above $\mathscr{I}_F$ and $\alpha > \beta$, then orbits approach $\mathscr{I}_F$ by spiralling toward Op. If the initial point is below $\mathscr{I}_F$ then orbits approach $\mathscr{I}_F$ by spiralling upwards toward Oq. The biological interpretation of these results is that in such a precariously balanced situation the long term outcome is coexistence of the two strains in proportions which depend on the initial configuration. This situation is unstable in the sense that a perturbation which disturbs the seed pool numbers at some instant, but which does not disturb the balance condition (5.25), will push the orbit to a neighboring one with a consequent change in long term abundances.

The geometric rate of convergence is perhaps a little surprising in view of (iii); attracting neutrally stable points are often approached at only an algebraically fast rate. For Model 1A the rate constant is

$$\gamma = \alpha p + \beta q/c;$$

recall

$$(p, q) \in \mathcal{I}_F \quad \text{iff} \quad cp + q = c.$$

Proof. Assertions (i) and (ii) can be demonstrated in much the same way as similar results above. For (iii), first observe that because of (5.25) and our basic assumptions, $zF(z)/(1 - z)$ is non-decreasing. It follows by differentiation that

$$\omega(z) = z(1 - z)F'(z)/F(z) \leqslant 1. \tag{5.29}$$

The Jacobian matrix (3.14) associated with the point $\mathbf{x} \in \mathcal{I}_F$ takes the form

$$\mathcal{J}(\mathbf{x}) = \begin{bmatrix} \alpha + (1 - \alpha)\omega(z) & -\dfrac{z(1 - \alpha)\omega(z)}{1 - z} \\ -(1 - z)(1 - \beta)(1 - \omega(z))/z & \beta + (1 - \beta)(1 - \omega(z)) \end{bmatrix}$$

and it is easily seen that the corresponding eigenvalues are $\lambda_1 = 1$ and $\lambda_2 = \alpha + (\beta - \alpha)\omega(z) < 1$.

To prove (5.28) recall that $F'(z) \geqslant 0$ and, from (3.15), $(p, q) \in \mathcal{I}_F$ iff $p = F(z)$. Then clearly $\gamma < \alpha$ if $\alpha > \beta$. If $\beta > \alpha$ then (5.29) yields

$$\gamma \leqslant \alpha + (\beta - \alpha)qF(z)/(p + q)^2 z(1 - z) = \beta,$$

and (5.28) follows. To prove the convergence assertions we can, without loss of generality, assume $(p_1, q_1) \in \mathcal{R}_3$, and set $\varepsilon_n = p_n - p$ and $\eta_n = q_n - q$. Then $\varepsilon_n, \eta_n \downarrow 0$, and $z_n \to p/(p + q)$ monotonically because, by (i), $\mathcal{R}_3$ coincides with either $\mathcal{R}_p$ or $\mathcal{R}_q$. With this notation we have

$$\varepsilon_{n+1} = \alpha\varepsilon_n + r_n \tag{5.30}$$

where

$$r_n = (1 - \alpha)(F(z_n) - F(z)).$$

The mean value theorem yields

$$r_n \sim (1 - \alpha) F'(z) \frac{q\varepsilon_n - p\eta_n}{(p_n + q_n)(p + q)} \tag{5.31}$$

and $q\varepsilon_n - p\eta_n$ has the same sign as r_n for all n. By expressing the left hand side of (5.27) in terms of ε_n and η_n and solving the resulting difference equation for η_n we obtain

$$\eta_n = \frac{1 - \beta}{1 - \alpha} \sum_{v=n}^{\infty} (q_v/p_v)(\varepsilon_v - \varepsilon_{v+1}) \sim [(1 - \beta)/(1 - \alpha)](q/p)\varepsilon_n.$$

It follows that

$$q\varepsilon_n - p\eta_n \sim [(\beta - \alpha)/(1 - \alpha)]q\varepsilon_n \tag{5.32}$$

and hence from (5.30) and (5.31) we obtain $\varepsilon_{n+1} \sim \gamma\varepsilon_n$. Consequently, if we set $L_n = \gamma^{-n}\varepsilon_n$, then $L_{n+1}/L_n \to 1$. In particular we can choose $\varepsilon > 0$ such that $\gamma(1 + \varepsilon)^2 < 1$ and then $n' > 0$ so that $0 < L_n \leqslant (1 + \varepsilon)^n$ when $n \geqslant n'$.

We now rewrite (5.30) as

$$L_{n+1} = L_n + \gamma^{-n-1} R_n$$

where

$$R_n = r_n - [(\beta - \alpha)F'(z)/(p + q)^2]q\varepsilon_n.$$

Applying MacLaurin's expansion to get a finer expansion of r_n and using (5.32) we obtain

$$r_n = (1 - \alpha) F'(z) \frac{q\varepsilon_n - p\eta_n}{(p_n + q_n)(p + q)} + F''(\xi_n)O(\varepsilon_n^2)$$

where ξ_n lies between z_n and z. It follows from the estimates obtained above that $\Sigma|F''(\xi_n)|\gamma^{-n}\varepsilon_n^2 < \infty$. The remaining contribution to R_n is

$$\gamma^{-n-1} \frac{F'(z)}{p + q} \left[\frac{1 - \alpha}{p_n + q_n}(q\varepsilon_n - p\eta_n) - \frac{\beta - \alpha}{p + q}q\varepsilon_n \right]$$

$$= \gamma^{-n-1} \left\{ \frac{F'(z)}{(p + q)(p_n + q_n)}[(1 - \beta)q\varepsilon_n - (1 - \alpha)p\eta_n] + O(\varepsilon_n^2) \right\}.$$

The term in square brackets is

$$(1 - \beta) \sum_{n=v}^{\infty} (q\varepsilon_v - p\eta_v)(\varepsilon_v - \varepsilon_{v+1})/p_v$$

$$\sim \left(\frac{1 - \beta}{1 - \alpha} \right) \left(\frac{q}{p} \right)(\beta - \alpha) \sum_{v=n}^{\infty} \varepsilon_v(\varepsilon_v - \varepsilon_{v+1}),$$

where we have used (5.32) and the fact that $q\varepsilon_n - p\eta_n$ has constant sign. Since the summands are bounded by $\varepsilon_v^2 - \varepsilon_{v+1}^2$, the whole term is $O(\varepsilon_n^2)$. It follows that $\Sigma \gamma^{-n} R_n$ converges and hence that $L_n \to L$ say, as asserted. This completes the proof.

6

Analysis of Model 2

6.1   Introduction

We now take account of the existence of free seed and the possibility of age dependent softening rates. For this general case we no longer have the simple relationship (3.1) between the year-to-year sizes of maximum seed pools. Instead we must try to work with the more complicated relationships governing the new seed productions given by Equations (2.25) and (2.26) for Strain 1. Although the assumptions made in Chapter 2 yielding Models 2A and 1 (but not Model 1A) simplify some of the parameters in a computational sense, they do not give simpler mathematical structures and hence in this chapter we will focus on Model 2 and comment no further on its special cases.

As for Model 1A, the time variation of P_V poses a problem for analysing the most general version of Model 2. In practice, however, the memory length N is always finite. Consequently we can use (2.26) with varying P_V to calculate $z_{n(V)}$ (where P_V is constant for $n > n(V)$) and then use this as an initial value for the system (2.26) with P_V set to its limiting value. Hence we will assume that P_V is a constant equal to its limiting value for all the analysis of this chapter.

Recall that $L = M_1 P_V$ and $M = M_2 P_V$, define

$$a_i = M_1 P_E P_A W S_i \quad \text{and} \quad b_i = M_2 Q_E Q_A T_i \quad (i \geqslant 1)$$

and, as in Chapters 3 and 4, let $c = k_{12}$ and $d = k_{21}$. It is convenient to standardize the new seed production numbers, x_n and y_n, by defining $X_n = x_n/L$ and $Y_n = y_n/M$. Our model equations (2.25) and (2.26) now take the form: for $n \geqslant 2$

$$X_n = \frac{c \sum_{i=1}^{n-1} a_i X_{n-i}}{c \sum_{i=1}^{n-1} a_i X_{n-i} + \sum_{i=1}^{n-1} b_i Y_{n-i}} \tag{6.1a}$$

and

$$Y_n = \frac{d \sum_{i=1}^{n-1} b_i Y_{n-i}}{\sum_{i=1}^{n-1} a_i X_{n-i} + d \sum_{i=1}^{n-1} b_i Y_{n-i}} \tag{6.1b}$$

where the initial conditions are simply that X_1 and Y_1 are arbitrary positive numbers.

As in Chapter 3 we could generalize this system by allowing $F(\cdot)$ and $G(\cdot)$ to be as defined therein, setting

$$Z_{n-1} = \sum_{i=1}^{n-1} b_i Y_{n-i} \bigg/ \sum_{i=1}^{n-1} (a_i X_{n-i} + b_i Y_{n-i}),$$

$X_n = F(Z_{n-1})$ and $Y_n = G(Z_{n-1})$. Since the system (6.1) provides enough difficulties of its own we will not pursue the analysis of this generalisation.

Strictly speaking, the system (6.1) is not a dynamical system in the sense that we defined this term in Chapter 3. If the a_i and/or the b_i are positive for infinitely many i then (6.1) has an infinite memory in the sense that we must know (X_1, Y_1), ..., (X_{n-1}, Y_{n-1}) to calculate (X_n, Y_n). Thus it is not possible to describe the state of the system at any given time by a vector in a Euclidean space of fixed dimension. In the botanically relevant finite memory case, $N < \infty$, this problem can be vanquished by embedding the system (6.1) into a dynamical system having a $2N$-dimensional Euclidean state space. We will see that this dynamical system can have an interior equilibrium which may be GAS. On the other hand, no matter whether N is finite or infinite, the system (6.1) will not in general possess interior equilibrium points. This can be seen as follows. If $X_2 = X_1$ and $Y_2 = Y_1$ then obviously X_1 and Y_1 are fully determined by a_1, b_1, c and d. Unless all the other a_i and b_i stand in some special relationship to these parameters we can expect X_n and Y_n for $n > 2$ to differ from X_1 and Y_1, respectively. It follows that the best result about (6.1) concerning a long-term co-existence of strains is that under appropriate conditions there will exist positive numbers X and Y such that $(X_n, Y_n) \to (X, Y)$ as $n \to \infty$. Using terminology introduced in Chapter 3, we will look for globally attracting points.

In this chapter we determine the limiting behavior of (6.1) when $N < \infty$ and we obtain a less complete result when $N = \infty$. Section 2 is concerned with the existence of boundary equilibria of system (6.1), details of the above-mentioned embedding, and the restriction property mentioned in Chapter 3. The de Wit case is considered in Section 3. When $N < \infty$ we show that typically the system converges to one of its boundary equilibria at a geometrically fast rate. Similar results can be given when $N = \infty$ but only under a rather strong dominance condition on the softening rates a_n and b_n; see Equation (6.11) below. An interior global attractor can exist only if $cd \neq 1$. A fairly thorough treatment of this case is given in Section 4 when $N < \infty$. We find conditions ensuring global attraction to one or other boundary equilibrium or to an interior point attractor, and we obtain results for the rate of convergence. When none of these conditions is satisfied there is a locally repelling interior point and two locally attracting boundary points. In this case it seems likely that the boundary equilibria have regions of attraction which essentially exhaust the state space. The best result in this direction is to the effect that if there is an N-long run of time points during which the system moves steadily toward one of the boundary equilibria then this motion continues indefinitely; see Theorem 6.12 below. We have found no way of obtaining global attraction results for the infinite memory model under general conditions. In Section 5 we present a local analysis under the regularity conditions $\Sigma a_n, \Sigma b_n < \infty$ which gives results in agreement with those derived in earlier sections. We end the chapter by looking at a couple of cases where these conditions are relaxed. These suggest that the global behaviors seen when Σa_n and Σb_n are infinite are only quantitatively changed in the sense that the approach to attracting points occurs more slowly.

6.2   Some preliminary results

Throughout this chapter we assume that S_1 and T_1 are positive (equivalently a_1 and b_1 are positive). The following result on Lagrange stability is then evident from (6.1).

Theorem 6.1.
If X_1 and $Y_1 > 0$ then for $n \geqslant 2$,

$$0 < X_n, \quad Y_n < 1.$$

We indicated in Chapter 3 that there is a simple relation between X_n and Y_n which facilitates the analysis of (6.1).

Theorem 6.2
For all $n \geqslant 2$ the points (X_n, Y_n) lie on the curve

$$y = \gamma(x) = cd(1 - x)/[cd - (cd - 1)x] \quad (0 < x < 1). \tag{6.2}$$

Proof. We will find it useful in the sequel to use the following notation. For $n \geqslant 2$ let $\Sigma_1 = \Sigma_{i=1}^{n-1} a_i X_{n-i}$ and $\Sigma_2 = \Sigma_{i=1}^{n-1} b_i Y_{n-i}$. Then (6.1) yields

$$c/X_n = c + \Sigma_2/\Sigma_1 \quad \text{and} \quad d/Y_n = d + \Sigma_1/\Sigma_2$$

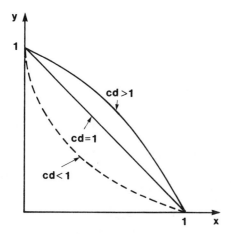

Figure 6.1. The motion of (X_n, Y_n) for Model 2.

and these can be combined into the symmetric form $cd(Y_n^{-1} - 1)$ $(X_n^{-1} - 1) = 1$. This is equivalent to the assertion.

Obviously the function defined by (6.2) can be extended to the domain $[0, 1]$ and we usually take $\gamma(\cdot)$ to be defined in this way. This function decreases from unity to zero as x increases through $[0, 1]$ and it is concave when $cd > 1$, linear when $cd = 1$ and convex when $cd < 1$; see Figure 6.1.

We see that the curve defined by (6.2), and denoted by $\mathscr{A}$, is an attracting set in the very strong sense that no matter where the system starts in $\mathbb{R}_+^2$ it jumps into $\mathscr{A}$ at $n = 2$ and stays there for all subsequent n. We will be much concerned with the question of which points in $\bar{\mathscr{A}}$ can be limit points of the motion.

Suppose that $X_1 = 0$ but $Y_1 > 0$. It is obvious that $\Sigma_1 + \Sigma_2 > 0$ but $X_2 = 0$, and hence by induction, for each $n \geqslant 2$, $X_n = 0$ and $Y_n > 0$. Consequently for all $n \geqslant 2$, $\Sigma_1 = 0$ and $\Sigma_2 > 0$ and (6.1b) shows that $Y_n = 1$. In particular if $(X_1, Y_1) = (0, 1)$ then $(X_n, Y_n) = (0, 1)$ and hence $(0, 1)$ is a boundary equilibrium point of the system (6.1). In addition if the system starts from any point $(0, Y_1)$ with $Y_1 > 0$ then it hits $(0, 1)$ at $n = 2$. The following result is now obvious.

Theorem 6.3.

The points $(0, 1)$ *and* $(1, 0)$ *are the unique boundary equilibria of* (6.1) *and* $(0, 1)$ *(respectively* $(1, 0)$*) is attained at* $n = 2$ *from any initial point* $(0, Y_1)$ *with* $Y_1 > 0$ *(respectively* $(X_1, 0)$ *with* $X_1 > 0$*).*

We now consider the finite memory case $N < \infty$. Observe that this does not contain Model 1A because there the sequences $\{a_i\}$ and $\{b_i\}$ are infinite geometric sequences. A dynamical system $\{\mathbf{Z}_n, n \geqslant 1\}$ with state space $\mathscr{S} = \mathbb{R}_+^{2N} \backslash \{0\}$ can be defined by the map $\mathscr{T} : \mathscr{S} \to \mathscr{S}$ where, if $\mathbf{Z} \in \mathscr{S}$ and Z_i denotes the ith component of $\mathbf{Z}$, then

$$(\mathscr{T}\mathbf{Z})_i = Z_{i+1} \quad \text{if} \quad i \neq N, 2N \tag{6.3a}$$

$$(\mathscr{T}\mathbf{Z})_N = c \sum_{i=1}^{N} a_i Z_{N-i+1} \bigg/ \sum_{i=1}^{N} [ca_i Z_{N-i+1} + b_i Z_{2N-i+1}] \tag{6.3b}$$

and

$$(\mathscr{T}\mathbf{Z})_{2N} = d \sum_{i=1}^{N} b_i Z_{2N-i+1} \bigg/ \sum_{i=1}^{N} [a_i Z_{N-i+1} + db_i Z_{2N-i+1}]. \tag{6.3c}$$

If we prescribe X_1, $Y_1 > 0$ and then set

$$\mathbf{Z}_n^T = (X_n, X_{n+1}, \ldots, X_{n+N-1}, Y_n, Y_{n+1}, \ldots, Y_{n+N-1}) \tag{6.4}$$

where the X_n and Y_n for $n \geq 2$ are obtained from (6.1) then it is not hard to show that $\mathbf{Z}_{n+1} = \mathscr{T}\mathbf{Z}_n$. Consequently any initial state of the system (6.1) determines an initial state $\mathbf{Z}_1$ of the dynamical system $\mathscr{T}$, and any $\mathbf{Z}_1 \in \mathscr{S}$ determines an orbit $\{\mathbf{Z}_n : n \in \mathbb{N}\}$ where $\mathbf{Z}_{n+1} = \mathscr{T}\mathbf{Z}_n$. This construction embeds the system (6.1) in the dynamical system (6.3). Clearly not all initial points $\mathbf{Z}_1 \in \mathscr{S}$ need be realised by the construction (6.4) when $n = 1$.

We now look for interior equilibria of $\mathscr{T}$ because they should suggest possible attractors for (6.1). We will always use the notation (even when $N = \infty$)

$$A = \sum_{i=1}^{N} a_i \quad \text{and} \quad B = \sum_{i=1}^{N} b_i.$$

Theorem 6.4.

If $cd \neq 1$ the dynamical system $\mathscr{T}$ has one interior equilibrium $\mathbf{Z}$ *iff*

$$1/d < B/A < c \; (implying \; cd > 1) \tag{6.5a}$$

or if

$$c < B/A < 1/d \; (implying \; cd < 1) \tag{6.5b}$$

and in either case it is given by

$$Z_i = X \quad if \quad 1 \leqslant i \leqslant N, = Y \quad if \quad N < i \leqslant 2N \tag{6.6}$$

where

$$X = \frac{d(cA - B)}{A(cd - 1)} \tag{6.7a}$$

and

$$Y = \frac{c(dB - A)}{B(cd - 1)}. \tag{6.7b}$$

If $cd = 1$ then interior equilibria exist iff

$$cA = B \qquad (6.8)$$

and then each point $\mathbf{Z} \in \mathscr{S}$ having the form (6.6) with $X + Y = 1$ is an equilibrium point.

Proof. By definition $\mathbf{Z}$ is an equilibrium iff $\mathscr{T}\mathbf{Z} = \mathbf{Z}$ and then from (6.3a) we have $Z_i = (\mathscr{T}\mathbf{Z})_i = Z_{i+1}$ if $i \neq N$, $2N$. Consequently $Z_1 = \ldots = Z_N$ and we denote these common values by X. Similarly $Z_{N+1} = \ldots = Z_{2N}$, which we denote by Y. It follows that $X = (\mathscr{T}\mathbf{Z})_N$ and hence (6.3b) yields

$$X = cAX/(cAX + BY).$$

Similarly

$$Y = dBY/(AX + dBY).$$

It follows that $\mathbf{Z}$ is an interior equilibrium iff both X and Y are positive and hence the last pair of equations reduce to linear equations. These have the solutions (6.7) iff $cd \neq 1$ and these solutions are positive iff either of (6.5) hold. If $cd = 1$ the linear equations reduce to $cAX + BY = cA = B$, and the final assertion follows.

The pair (X, Y) is an equilibrium point of the system (6.3) only if

$$X = cX\hat{a}_n/(cX\hat{a}_n + Y\hat{b}_n) \quad (n \geq 1)$$

where $(\cdot)_n$ denotes summation of n terms, for example $\hat{a}_n = \Sigma_{j=1}^n a_j$. This condition and the similar one derived from (6.3b) can hold iff $\hat{a}_n/\hat{b}_n \equiv A/B$, a condition which we would not normally expect. However, Theorem 6.4 identifies (X, Y) as a possible globally attracting point provided (6.5) holds. Whether or not this point really is globally attracting can only be determined by further analysis.

We can search for boundary equilibria of $\mathscr{T}$ by setting, for example, $Z_i = 0$ for some $1 \leq i \leq N$. This implies that $Z_1 = \ldots = Z_N = 0$ and then $\mathscr{T}\mathbf{Z} = \mathbf{Z}$ can be solved by taking $Z_{N+1} = \ldots = Z_{2N} = 1$. In this way we find that there are exactly two boundary equilibria and these have the form (6.6) with $X = 0$ and $Y = 1$, and $X = 1$ and $Y = 0$, respectively. Clearly, if $\{\mathscr{T}^n\mathbf{Z}\}$ converges for all $\mathbf{Z} \in \mathscr{S}^0$ to one of the three equilibria we have identified, then $\{(X_n, Y_n)\}$ will converge for all $X_1, Y_1 > 0$ to the appropriate limit.

6.3 Attraction of equilibria when $cd = 1$

We begin our discussion by considering the important de Wit case defined by $cd = 1$. If $cA \neq B$ we have seen that the system (6.3) possesses two boundary equilibria and there is no obvious candidate for an internal globally attracting point. The following theorem shows that precisely one of the boundary equilibria attract all interior points.

> *Theorem 6.5*
> *Suppose* $cd = 1$, $N < \infty$ *and* $0 < X_1$, $Y_1 < \infty$.

(i) *If* $cA < B$ *then*

$$X_n = 1 - Y_n = O((cA/B)^{n/N}).$$

(ii) *If* $cA > B$ *then*

$$Y_n = 1 - X_n = O((B/cA)^{n/N}).$$

Proof. Assume that the condition under (i) holds. Since $N < \infty$ it follows from Theorem 6.2 that for $m \geqslant 2$, $X_{m+N} = \phi(X_{m+N-1}, \ldots, X_m)$ where for $0 < u_1, \ldots, u_N < 1$ we define

$$\phi(u_1, \ldots, u_N) = c \sum_{i=1}^N a_i u_i \bigg/ \sum_{i=1}^N (ca_i u_i + b_i(1 - u_i)) \qquad (6.9)$$

and $X_2, \ldots, X_N$ are determined by (6.1) and prescribed X_1 and Y_1. Some algebraic manipulation shows that $\partial\phi/\partial u_j > 0$ in $(0, 1)^N$.

For a fixed integer $m \geqslant 2$ we can choose $0 < \varepsilon < 1$ such that $0 < X_m$, $\ldots, X_{m+N-1} < \varepsilon$. This follows because the system (6.1) cannot hit the co-ordinate axes in a finite number of steps. Consequently

$$X_{m+N} < \phi(\varepsilon, \ldots, \varepsilon).$$

If $f(\varepsilon)$ denotes the upper bound then clearly

$$f(s) = cAs/[B - (B - cA)s]$$

and this is the generating function of the geometric distribution which attributes probability $(cA/B)(1 - cA/B)^{j-1}$ to the point $j \in \mathbb{N}$. Since $\varepsilon < 1$ we have $f(\varepsilon)/\varepsilon < 1$, whence $X_{m+N} < \varepsilon$. Thus $0 < X_{m+1}, \ldots, X_{m+n} < \varepsilon$ and repetition of this argument shows that $X_{m+N+1} < f(\varepsilon)$. By arguing inductively it is clear that

$$X_{m+N}, \ldots, X_{m+2N-1} < f(\varepsilon).$$

Repetition of these arguments with ε replaced by $f(\varepsilon)$ shows that

$$X_{m+2N}, \ldots, X_{m+3N-1} < f(f(\varepsilon)).$$

Let $f_n(s)$ denote the n-fold composition of $f(s)$ which is defined by $f_0(s) = s$ and $f_n(s) = f(f_{n-1}(s))$ $(n \geqslant 1)$. A continuation of the above arguments yields, for $k \in \mathbb{N}$,

$$X_{m+kN}, \ldots, X_{m+(k+1)N-1} < f_k(\varepsilon). \qquad (6.10)$$

The theory of functional iteration (Harris (1963), p. 8) shows that $f_k(\varepsilon) \to 0$ as $k \to \infty$ and we conclude that $X_n \to 0$, and also, since $Y_n = 1 - X_n$, that $Y_n \to 1$. This shows that the point $(0, 1)$ is globally attracting for interior starting points.

The function $f(\varepsilon)$ satisfies the conditions of a theorem of Szekeres (see Kuczma (1968), p. 138) which states that $\lim_{n \to \infty} \theta^{-n} f_n(\varepsilon)$ exists as a positive number, where

$$\theta = f'(0) = cA/B.$$

This completes the proof of (i). The proof of (ii) is almost identical. In this case we work with $\{Y_n\}$ using (6.1b) and expressing the inequality hypothesis as $dB < A$.

We can reformulate this theorem in terms of the original model parameters as follows. The de Wit condition is $k_{12}k_{21} = 1$. The inequality condition under (i) becomes

$$k_{12} P_E P_A W \sum_{i=1}^{N} S_i < Q_E Q_A \sum_{i=1}^{N} T_i.$$

Recall that P_V will typically depend on n. Assuming that $\lambda_V = \lim_{n \to \infty} P_V$ exists and is positive, the theorem states that Strain 1 is driven to extinction and Strain 2 persists with the limiting new seed production $\lambda_V M_2$. Conversely, if the above inequality is reversed then Strain 2 is driven to extinction and Strain 1 persists with the limiting new seed production $\lambda_V M_1$.

We now consider the infinite memory case $N = \infty$. This seems to be botanically much less relevant than the finite memory case because it assumes a positive rate of softening no matter how old is the seed. On the other hand, this assumption is incorporated into the useful Model 1A, but then the form of the softening rates is such as to allow the rather complete analysis of previous chapters. The general infinite memory case is much less tractable and to prove the convergence assertions of Theorem 6.5 we find it necessary to strengthen the hypotheses used there.

Theorem 6.6.
Suppose $cd = 1$, $N = \infty$, X_1, $Y_1 > 0$, $A < \infty$, $cA < B$ and for $n \geqslant 1$,

$$ca_n \leqslant b_n. \tag{6.11}$$

Then $\lim_{n \to \infty} X_n = 0$ and $\lim_{n \to \infty} Y_n = 1$. On the other hand, if $B < \infty$, $cA > B$ and $ca_n \geqslant b_n$, $(n \geqslant 1)$ then $\lim_{n \to \infty} X_n = 1$ and $\lim_{n \to \infty} Y_n = 0$.

Proof. We prove only the direct assertion. By using Theorem 6.2 we can construct the sequence $\{X_n\}$ in the following manner. For $n \geqslant 2$ define the functions

$$\phi_{n-1}(u_1, \ldots, u_{n-1}; v) = \frac{c \sum_{i=1}^{n-1} a_i u_i}{ca_{n-1}u_{n-1} + b_{n-1}v + \hat{b}_{n-2} + \sum_{i=1}^{n-2} (ca_i - b_i)u_i}$$

where $0 < u_i < 1$, $(1 \leqslant i \leqslant n - 2)$ and $u_{n-1}, v > 0$. It is clear that

$$X_n = \phi_{n-1}(X_{n-1}, \ldots, X_1; Y_1).$$

The next step is to show that there exists $0 < \varepsilon < 1$ such that $X_n \leqslant \varepsilon$ when $n \geqslant 2$. Observe that $\phi_{n-1}(\cdot)$ is non-decreasing in $u_2, \ldots, u_{n-1}$ and for a given n there exists $\varepsilon \in (1/2, 1)$ such that $X_2, \ldots, X_{n-1} < \varepsilon$. Consequently

$$X_n < \frac{c(a_{n-1}X_1 + \varepsilon\hat{a}_{n-2})}{ca_{n-1}X_1 + b_{n-1}Y_1 + \hat{b}_{n-2} + \varepsilon(c\hat{a}_{n-2} - \hat{b}_{n-2})}$$

$$= \frac{c(a_{n-1}X_1 + \varepsilon\hat{a}_{n-2})}{c(a_{n-1}X_1 + \varepsilon\hat{a}_{n-2}) + b_{n-1}Y_1 + \hat{b}_{n-2}(1 - \varepsilon)}.$$

It follows that

$$\varepsilon - X_n > \frac{\varepsilon(1 - \varepsilon)(\hat{b}_{n-2} - c\hat{a}_{n-2}) + \varepsilon b_{n-1}Y_1 - (1 - \varepsilon)ca_{n-1}X_1}{c(a_{n-1}X_1 + \varepsilon\hat{a}_{n-2}) + b_{n-1}Y_1 + \hat{b}_{n-2}(1 - \varepsilon)}.$$

But $a_{n-1} \to 0$ and hence there exists n_0 such that for all $n \geqslant n_0$, $\hat{b}_{n-2} - c\hat{a}_{n-2} > (B - cA)/2 > 0$. It follows, by increasing n_0 if necessary, that for $n \geqslant n_0$ we have

$$(\hat{b}_{n-2} - c\hat{a}_{n-2})/2 - ca_{n-1}X_1 > 0.$$

Since $\varepsilon > 1/2$ we can make our initial choice of ε and n large enough to conclude that $X_1, \ldots, X_{n-1} < \varepsilon$ implies that $X_n < \varepsilon$. An induction argument now shows that $X_m < \varepsilon$ for all $m \geqslant 2$.

With n and ε fixed as above, (6.11) implies that

$$X_m < c \sum_{i=1}^{m-1} a_i X_{m-i} \bigg/ \left[\hat{b}_{m-2} - \sum_{i=1}^{m-2} (b_i - ca_i)X_{m-i} \right]$$

$$\leqslant c \sum_{i=1}^{m-1} a_i X_{m-i} / [c\varepsilon\hat{a}_{m-2} + \hat{b}_{m-2}(1 - \varepsilon)].$$

We can write this as

$$X_m < \tau_m + \rho \sum_{i=1}^{m-1} a_i X_{m-i}$$

where

$$\rho = cA/(\varepsilon cA + (1 - \varepsilon)B) < 1$$

and

$$\tau_m = [(\varepsilon c \hat{a}_{m-2} + (1 - \varepsilon)\hat{b}_{m-2})^{-1} - (\varepsilon cA + (1 - \varepsilon)B)^{-1}]c \sum_{i=1}^{m-1} a_i X_{m-i}.$$

Since $\{X_m\}$ is bounded it follows that $\tau_m \to 0$ and hence $\lim_{m\to\infty} \sup X_m \leqslant \rho \lim \sup_{m\to\infty} X_m$. Consequently $\lim \sup_{m\to\infty} X_m = 0$, and the proof is complete.

Just as for Theorem 6.5, this theorem can also be reformulated in terms of the original model parameters. The proof of Theorem 6.6 gives no information on how rapidly the limits are approached. Our next results show that convergence is geometrically fast and it gives the exact rate, thus refining the estimates given in Theorem 6.5, but under a further assumption.

Theorem 6.7.
Let $cd = 1$, X_1, $Y_1 > 0$, $cA < B$ and either (i) $N < \infty$; or (ii) $N = \infty$ and (6.11) holds. In addition suppose the following conditions are satisfied:

$$\text{g.c.d. } \{n : a_n > 0\} = 1; \tag{6.12a}$$

there exists $\eta > 1$ such that

$$(c/B) \sum_{n=1}^{\infty} a_n \eta^n = 1; \tag{6.12b}$$

$$\sum_{n=1}^{\infty} n a_n \eta^n < \infty; \tag{(6.12c)}$$

and there exists $\xi \in (1, \eta)$ such that

$$\sum_{n \geqslant 1} b_n \xi^n < \infty. \tag{6.12d}$$

Then $\lim_{n\to\infty} \eta^n X_n$ exists and is finite.

Remarks. It is likely that (6.11) is not necessary for this result. When $N < \infty$ the equation determining η is a polynomial and hence is relatively easy to solve. Theorem 6.7 is the best possible result in the sense that if $\eta' > \eta$ then $(\eta')^n X_n \to \infty$. Our proof does not preclude the possibility that the limit is zero. We will comment on this following the proof. Finally, observe that the hypotheses entail the finitude of A and B.

Proof. The proof proceeds in several stages. First we show that if ζ is any constant satisfying $1 < \zeta < \eta$ then $\lambda_n = \zeta^n X_n \to 0$ as $n \to \infty$. Write (6.1a) as

$$X_n = (1 + \delta_n)(c/B) \sum_{i=1}^{n-1} a_i X_{n-i} \tag{6.13}$$

where

$$\delta_n = \frac{\check{b}_{n-1} - b_{n-1}(X_1 + Y_1) + \sum_{i=1}^{n-1} (b_i - a_i)X_{n-i}}{\sum_{i=1}^{n-1} (ca_i X_{n-i} + b_i Y_{n-i})}$$

and $\check{b}_n = B - \check{b}_{n-1}$. Let γ be a constant and $\{\phi_n : n \geqslant 1\}$ be a sequence satisfying $\gamma\phi_n = (c/B)\zeta^n a_n$ and $\sum_{n \geqslant 1} \phi_n = 1$. Then (6.13) can be re-written as

$$\lambda_n = (1 + \delta_n)\gamma \sum_{i=1}^{n-1} \phi_i \lambda_{n-i}.$$

The definition of η shows that $\gamma < 1$ and Theorems 6.5 and 6.6 imply that $\delta_n \to 0$. Thus we can find $\rho < 1$ and n' such that $(1 + \delta_n)\gamma < \rho$ when $n \geqslant n'$. For such n,

$$\lambda_n = \rho \sum_{i=1}^{n-1} \phi_i \lambda_{n-i} + [(1 + \delta_n)\gamma - \rho] \sum_{i=1}^{n-1} \phi_i \lambda_{n-i}$$

where the term in square brackets is negative. It follows then that there is a positive constant K such that

$$\lambda_{n+1} \leqslant \rho \sum_{i=0}^{n} \phi_i \lambda_{n+1-i} + K_n$$

where $K_n = K$ if $n < n'$, $= 0$ if $n \geqslant n'$. Let $\{\phi_i^{(n)} : i \geqslant 0\}$ denote the n-fold convolution of the probability distribution $\{\phi_i\}$. Iteration of the above inequality shows that for each $v \in \mathbb{N}_+$,

$$\lambda_{n+1} \leqslant \rho^{v+1} \sum_{i=0}^{n} \phi_i^{(v+1)} \lambda_{n+1-i} + \sum_{j=0}^{v} \rho^j \sum_{i=0}^{n} K_i \phi_{n-i}^{(j)}.$$

The first term on the right tends to zero as $v \to \infty$, whence

$$\lambda_{n+1} \leqslant K \sum_{j=0}^{\infty} \rho^j \sum_{i=0}^{n'} \phi_{n-i}^{(j)}$$

and dominated convergence shows that the right hand side tends to zero as $n \to \infty$.

To complete the proof, let $\Lambda_n = \eta^n X_n$ and $p_n = (c/B)\eta^n a_n$. We see from (6.12b) that $\{p_n : n \geqslant 1\}$ is a probability distribution. Equation

(6.13) yields the discrete renewal equation

$$\Lambda_n = \tau_n + \sum_{i=1}^{n-1} p_i \Lambda_{n-i}$$

where $\tau_n = (c/\hat{b}_n)\eta^n \delta_n \sum_{i=1}^{n-1} a_i X_{n-i}$. With ξ specified as in (6.12d), choose $1 < \theta < \eta$ such that $\eta/\xi\theta < 1$. Then

$$\tau_n = \frac{c}{B} \left\{ \frac{\xi^n \check{b}_{n-1} - \xi^n b_{n-1}(X_1 + Y_1) + \sum_{i=1}^{n-1} \xi^i (b_i - ca_i)\xi^{n-i} X_{n-i}}{\sum_{i=1}^{n-1} (ca_i X_{n-i} + b_i Y_{n-i})} \right\}$$

$$\times (\eta/\xi\theta)^n \sum_{i=1}^{n-1} \theta^i a_i \theta^{n-i} X_{n-i}.$$

The first stage of the proof shows that $\theta^n X_n \to 0$ and it follows from the dominated convergence theorem and (6.12d) that $|\tau_n| = O((\eta/\xi\theta)^n)$. In particular $\Sigma |\tau_n| < \infty$ and this condition, together with (6.12a) and (6.12c) allows us to apply the key renewal theorem (Feller (1971), p. 363) and conclude that

$$\Lambda_n \to \left(\sum_{n=1}^{\infty} \tau_n \right) \bigg/ \sum_{n=1}^{\infty} n p_n$$

which is finite and non-negative. This completes the proof.

It seems likely that $\sum_{n=1}^{\infty} \tau_n > 0$, but it is not clear how to establish this. The sign of τ_n is the same as that of δ_n and it is not difficult to show this can be made negative for fixed values of n by choosing X_1 and/or Y_1 to be sufficiently large. On the other hand, when $N < \infty$ and (6.11) holds then $\tau_n > 0$ when $n > N$.

A dual version of Theorem 6.7 obviously applies when $cA > B$. In this case we require the existence of $\tilde{\eta} > 1$ such that

$$(d/A) \sum_{n=1}^{\infty} b_n \tilde{\eta}^n = 1$$

and then under obvious further conditions $\lim_{n \to \infty} \tilde{\eta}^n Y_n$ exists and is finite.

It is of some interest to apply the results of Theorems 6.6 and 6.7 to Model 1A. Using the notation introduced at the beginning of Chapter 3 it follows from (2.14) that for $i \geqslant 1$,

$$a_i = M_1 X \alpha^{i-1} \quad \text{and} \quad b_i = M_2 Y \beta^{i-1}.$$

These yield $A = M_1 X/(1 - \alpha)$ and $B = M_2 Y/(1 - \beta)$ and hence $\rho = B/A$. Thus the stability condition $cA < B$ in Theorem 6.6 is just the condition

$c < \rho$ we encountered in Section 4.3 (I). The additional conditions (6.11) need not be satisfied by Model 1A and, as we know, they are not required for the stability results in Chapter 4.

Turning to Theorem 6.7, it is obvious that (6.12a) is satisfied and the equation defining η, (6.12b), has the solution $\eta = 1/\Delta$ where $\Delta = (1 - \alpha)c/\rho + \alpha$. The latter quantity was defined for Model G in Theorem 4.2. Clearly $\alpha\eta < 1$ so (6.12c) is satisfied and (6.12d) can be satisfied by choosing $\xi \in (1, 1/\beta)$. Recalling that $X_n = x_n/L$ and $x_n = t_n - \alpha t_{n-1}$, we see that the convergence rate results of Theorems 4.2 and 6.7 agree for Model 1A, as indeed they should.

We will end this section by looking at the case where $cd = 1$ and $cA = B$ which is not covered by Theorems 6.5 or 6.6. The last assertion of Theorem 6.4 suggests that the set $\{(X, Y) : X + Y = 1, 0 < X < 1\}$ is a minimal global attractor for (6.1) in the sense that each point in this set is the limit point of some orbit. We will see that orbits converge to this set when $N < \infty$, but when $N = \infty$ we need to impose rather severe supplementary conditions to obtain any results. These are not required for Model 1A where, with our present assumptions, the linear isoclines coincide.

Theorem 6.8.
If $cd = 1$, $N < \infty$ and $cA = B$ then for $n \geq 2$, $0 < \min(X_2, \ldots, X_{N+1}) \leq X_n \leq \max(X_2, \ldots, X_{N+1}) < 1$.

If also $a_n + b_n > 0$ for $n \leq N$ then $\{X_n\}$ and $\{Y_n\}$ have limits, X and Y, respectively, satisfying $X + Y = 1$ and $0 < X < 1$.

Proof. If Δ_1 denotes the above maximum, then $\Delta_1 < 1$. If $0 \leq x \leq 1$ it follows from (6.9) that $\phi(x, \ldots, x) = x$ and hence, since ϕ is non-decreasing in each of its arguments, $X_{N+2} \leq \Delta_1$. By proceeding inductively we obtain $X_n \leq \Delta_1$, $(n \geq 2)$, and in the same way we find that $X_n \geq \delta_1$, $(n \geq 2)$ where δ_1 is the above minimum, which obviously is positive. This proves the first assertion.

For $k \in \mathbb{N}$, let $\Delta_k = \max\{X_{2+(k-1)N}, \ldots, X_{1+kN}\}$. Since $X_n \leq \Delta_1$ it follows that $\Delta_2 \leq \Delta_1$ whence, by induction, $\{\Delta_k\}$ is non-increasing. Similarly, if $\delta_k = \min\{X_{2+(k-1)N}, \ldots, X_{1+kN}\}$, then $\{\delta_k\}$ is non-decreasing. Now either there exists $v \geq 2$ such that $X_v = \ldots = X_{v+N-1}$, or there is no such v. In the former case $X_n = X_v$ if $n \geq v$ and then $\lim_{n \to \infty} X_n$ exists.

Suppose now there is no such v. The condition $a_n + b_n > 0$ ensures that ϕ is strictly increasing in each of its arguments. Since not all of $X_{2+(k-1)N}, \ldots, X_{1+kN}$ are equal, it follows that $\delta_k < X_n < \Delta_k$ for $n = 2 + kN, \ldots, 1 + (k + 1)N$ and hence $\{\Delta_k\}$ and $\{\delta_k\}$ are strictly decreasing and increasing, respectively. Being bounded both sequences have limits which we denote by Δ and δ, respectively. Suppose that $\delta < \Delta$. There must then be sub-sequences of $\{X_n\}$ which converge to Δ and δ and because of our definition of Δ_k and δ_k, given $\varepsilon \in (0, (\Delta - \delta)/2)$, it is the case that for all n large enough there is one element of $\{X_n, \ldots, X_{n+N-1}\}$ in $(\Delta - \varepsilon, \Delta + \varepsilon)$ and another in $(\delta - \varepsilon, \delta + \varepsilon)$. We will now demonstrate the impossibility of this configuration and hence conclude that $\Delta = \delta$.

Choose k so large that the following conditions are satisfied:

(a) For $n > kN$, $\delta - \varepsilon \leqslant X_n \leqslant \Delta + \varepsilon$;
(b) There is an element of $\{X_{1+(k+1)N}, \ldots, X_{(k+2)N}\}$, X_j say, such that $\Delta - \varepsilon \leqslant X_j \leqslant \Delta + \varepsilon$; and
(c) The set $\{X_{j-N}, \ldots, X_{j-1}\}$ contains an element η in $[\delta - \varepsilon, \delta + \varepsilon]$.

How small can η be? Let m denote the subscript of the coefficients of η in the expression $X_j = \phi(X_{j-1}, \ldots, X_{j-N})$. Because of the monotone nature of ϕ, η can take its least value if $\Delta_j = \Delta - \varepsilon$ and all elements among $X_{j-N}, \ldots, X_{j-1}$, apart from η, equal $\Delta + \varepsilon$. Solving the equation $\Delta - \varepsilon = \phi(\Delta + \varepsilon, \ldots, \Delta + \varepsilon, \eta, \Delta + \varepsilon \ldots)$ we see that

$$\eta \geqslant \eta_{\min} = \Delta - \varepsilon \frac{c(2\Delta - a_m) - (\Delta - \varepsilon)(b_m - ca_m)}{ca_m + (\Delta - \varepsilon)(b_m - ca_m)}.$$

In particular $\eta_{\min}$ does not depend on δ and hence we can make an initial choice of ε so small that $\eta_{\min} > \delta + \varepsilon$. This is the desired contradiction and hence we conclude that $X_n \to X \in (0, 1)$ and $Y_n = 1 - X_n \to 1 - X$.

In the next theorem we admit the infinite memory case but impose the very strong balance condition (6.14a) below. We will see quite clearly that the limit point of each point depends on its initial point.

Theorem 6.9.

Let $cd = 1$, $N \leqslant \infty$, and suppose the following conditions are satisfied:

g.c.d. $\{n : a_n > 0\} = 1$;

$ca_n = b_n$ *when* $n \geqslant 1$; *and*   (6.14a)

$\sum na_n < \infty.$ (6.14b)

Then $\{X_n\}$ and $\{Y_n\}$ converge to limits, X and Y respectively, satisfying $X + Y = 1$ and depending on the initial conditions.

Proof. Theorem 6.2 says that $X_n + Y_n = 1$ if $n \geq 2$ whence from (6.13),

$$\sum_{i=1}^{n-1} (ca_i X_{n-i} + b_i Y_{n-i}) = ca_{n-1}(X_1 + Y_1) + c\hat{a}_{n-2}.$$

Let $\alpha_n = a_n/A$. Then (6.1a) can be written as the renewal equation

$$X_n = \tau_n + \sum_{i=1}^{n-1} \alpha_i X_{n-i}$$

where, if $\breve{a}_n = A - \hat{a}_{n-1}$,

$$\tau_n = \frac{\breve{a}_{n-1} - a_{n-1}(X_1 + Y_1)}{\hat{a}_{n-2} + a_{n-1}(X_1 + Y_1)} \sum_{i=1}^{n-1} \alpha_i X_{n-i}.$$

Our assumptions imply that $\{\alpha_i\}$ is an aperiodic discrete distribution and clearly $\tau_n = O(\breve{a}_{n-1})$. It follows from (6.14b) that $\Sigma |\tau_n| < \infty$ and hence we can apply the key renewal theorem to conclude that

$$\lim_{n \to \infty} X_n = \left(\sum_{n=1}^{\infty} \tau_n \right) \Big/ \sum_{n=1}^{\infty} n\alpha_n.$$

6.4 Long-term behavior when $cd \neq 1$, but $N < \infty$

The case $cd \neq 1$ is less easy to analyse than the de Wit case examined in the previous section. We will see in this section that the methodology used to prove Theorem 6.5 can be adapted to cope with the finite memory case. First, we consider those cases where the system (6.3) has no interior equilibrium, and then exactly one of the boundary equilibria is globally attracting.

Theorem 6.10.

When $cd \neq 1$, $N < \infty$ and the system (6.3) has no interior equilibrium then exactly one of the following cases will obtain:

(i) *If X_1, $Y_1 > 0$ then $X_n \to 0$ and $Y_n \to 1$ if either*

$$cd > 1 \text{ and } cA \leq B \qquad (6.15)$$

or

$$cd < 1 \text{ and } A \leq dB \qquad (6.16)$$

(ii) *If X_1, $Y_1 > 0$ then $X_n \to 1$ and $Y_n \to 0$ if either*

$$cd > 1 \text{ and } dB \leq A \qquad (6.17)$$

or

$$cd < 1 \text{ and } B \leq cA. \qquad (6.18)$$

In all cases convergence occurs geometrically fast when strict inequality holds, but only like $1/n$ in the case of equality.

Remark. More precisely we will show that when strict inequality holds in (6.15) then X_n and $1 - Y_n$ are $O(cA/B)^{n/N}$. Analogous assertions are true for the strict versions of (6.16)–(6.18).

Proof. We begin by extending the definition of ϕ given by (6.9). Define $\phi : \mathbb{R}^N \to [0, 1]$ by

$$\phi(u_1, \ldots, u_N) = c \sum_{i=1}^N a_i u_i \Bigg/ \left[c \sum_{i=1}^N a_i u_i + \sum_{i=1}^N b_i v_i \right] \tag{6.19}$$

where $v_i = \gamma(u_i)$ and γ is defined by (6.2). It is not hard to see that when $m \in \mathbb{N}$ we have

$$X_{m+N} = \phi(X_{m+N-1}, \ldots, X_m)$$

and $X_2, \ldots, X_N$ are determined from X_1 and Y_1 by (6.1). No matter what is the value of cd, we always have $\gamma'(x) < 0$ if $0 < x < 1$, and if Σ_1 and Σ_2 denote the sums in the u_i and the v_i occurring in (6.19), respectively, then

$$\partial\phi/\partial u_j = c[a_j \Sigma_2 - b_j \Sigma_1 \gamma'(u_j)]/[c\Sigma_1 + \Sigma_2]^2 \geq 0.$$

It follows that for any parameter combination, ϕ is non-decreasing in each of its arguments, and it is strictly increasing in at least one component, for if not then $a_j + b_j = 0$ for $1 \leq j \leq N$. Assume now that (6.15) holds. By arguing as in the proof of Theorem 6.5 we find that for a fixed $m \geq 2$ there exists $\varepsilon \in (0, 1)$ such that $X_{m+N} < f(\varepsilon)$ where $f(\varepsilon) = \phi(\varepsilon, \ldots, \varepsilon)$ is given by

$$f(s) = \frac{rs(1 - Qs)}{rs(1 - Qs) + 1 - s} \quad (0 \leq s \leq 1), \tag{6.20}$$

where $Q = 1 - 1/cd < 1$ and $r = cA/B \leq 1$. Obviously $f(0) = 0, f(1) = 1$ and $0 < f(s) < 1$ if $0 < s < 1$, but now $f(\cdot)$ may not be the generating function of any probability distribution. However

$$f'(s) = \frac{r[1 - 2Qs + Qs^2]}{[rs(1 - Qs) + 1 - s]^2} \tag{6.21}$$

and this is positive if $0 \leq s \leq 1$. Finally, it is easy to show that $f(s) < s$ if $0 < s < 1$. Under these conditions it is still true that $f_k(s) \to 0$ as $k \to \infty$ when $s < 1$; see Theorem 0.4 on p. 21 in Kuczma (1968). These properties of f allow us to repeat the argument leading to (6.10) and we conclude that $X_n \to 0$. It follows then from (6.2) that $Y_n \to 1$.

When $r < 1$, which corresponds to strict inequality in (6.15), Szekeres's theorem is applicable and, as in Theorem 6.5, we have

$X_n = O(r^{n/N})$. When $r = 1$ the maximal convergence rate is much slower than this geometric rate. We now have $f'(0) = 1, f''(0) = -2Q$, whence $f'(s) - 1 + 2Qs = O(s^2)$ as $s \to 0$. This condition allows us to invoke a result arising in the course of proving another theorem of Szekeres (see Equation (7.24) on p. 169 of Kuczma (1968)) and it states that $\lim_{k \to \infty} k f_k(s) = 2Q$ if $0 < s < 1$. We conclude that $X_n = O(1/n)$ as $n \to \infty$.

We now consider the parameter combinations in Theorem 6.4 entailing the existence of the interior equilibrium given by (6.6) and (6.7). Our next result shows when (X, Y) is globally attracting for the system (6.1).

> *Theorem 6.11.*
> *Suppose X_1, $Y_1 > 0$, $N < \infty$, and that (6.5a) is satisfied. Then*
>
> $$\theta = 2 - r + (r - 1)^2/rQ < 1$$
>
> *and*
>
> $$X_n - X = O(\theta^{n/N}) = Y_n - Y.$$

Proof. Condition (6.5a) can be expressed in the form

$$r(1 - Q) < 1 < r. \tag{6.22}$$

We use f as defined by (6.20) and observe that the equation $f(s) = s$ has exactly three solutions, namely, $s = 0$, $s = X$, and $s = 1$. Indeed, since $X = (r - 1)/Qr$, we have

$$f(s) - s = \frac{sQr(1 - s)(s - X)}{rs(1 - Qs) + 1 - s} \tag{6.23}$$

whence $f(s) < s$ if $s < X$ and $f(s) > s$ when $s > X$. If we proceed as in the proof of Theorem 6.5, making sure that $\varepsilon > X$, then the second condition above allows us to derive (6.10) once again. On the other hand, by choosing $\delta \in (0, X)$ so that $\delta < X_m, \ldots, X_{m+N-1}$, we can prove that

$$f_k(\delta) < X_{m+kN}, \ldots, X_{m+(k+1)N-1}.$$

The above properties of f ensure that $f_k(s) \to X$ when $0 < s < 1$ and hence $X_n \to X$; see Kuczma (1968), Theorem 0.4. Furthermore Szekeres's theorem says that

$$f_k(s) - X \sim \text{const.} \, (f'(X))^k.$$

It is easily shown that $\theta = f'(X)$ and from (6.22),

$$1 - \theta = (r - 1)(1 - r(1 - Q))/rQ > 0.$$

This completes the proof.

The only case not yet examined is that described by (6.5b). In the next section we analyse a linearized version of (6.1) which shows that when (6.5b) holds, both the boundary equilibria are locally attracting and (X, Y) is locally repelling, that is, orbits of the linearized system starting near (X, Y) diverge from it. This suggests that once an orbit hits the attracting curve $\mathscr{A}$ then it hops about and eventually converges to one of the boundary equilibria. The position of the initial point determines which of the equilibria attracts the orbit. Even though we are unable to prove that this *is* the case, we can give two results to the effect that if a certain regularity can be seen in some N-long segment of the orbit then the final outcome can be predicted. The first such result is not tied to any specific conditions on the parameters.

Theorem 6.12.
If $N < \infty$ and there exists $m > 1$ such that

$$X_m \leqslant X_{m+1} \leqslant \ldots \leqslant X_{m+N}, \tag{6.24}$$

then $X_n \leqslant X_{n+1}$ ($n \geqslant m$). This remains true if the inequalities are reversed and/or the Y_ns are substituted for the X_ns.

Proof. We saw in the proof of Theorem 6.10 that ϕ is non-decreasing in each of its arguments and hence it follows from (6.24) that

$$X_{m+N} = \phi(X_{m+N-1}, \ldots, X_m) \leqslant \phi(X_{m+N}, \ldots, X_{m+1}) = X_{m+N+1}.$$

The assertion follows by induction.

Numerical examples suggest that orbits jump haphazardly along $\mathscr{A}$, but eventually a condition like (6.24) is satisfied. Such orbits continue moving monotonically toward an attractor, that is, one of the boundary equilibria or (X, Y). This insight is directly applicable to Theorems 6.10 and 6.11. It also admits the possibility that, if in addition to (6.24) we have $X_{m+N} < X$ and (6.5b), then $X_n \to X$. The next theorem precludes this possibility.

Theorem 6.13.
Suppose $N < \infty$ and (6.5b) holds. If there exists $m \geqslant 2$ such that

$$X_m, \ldots, X_{m+N-1} < X \tag{6.25}$$

then $X_n \to 0$ as $n \to \infty$. More precisely, $r < 1$ and

$$X_n = O(r^{n/N}).$$

Conversely, if $X_m, \ldots, X_{m+N-1} > X$ then $X_n \to 1$, and, more precisely, $r(1 - Q) > 1$ and

$$1 - X_n = O[(r(1 - Q))^{-n/N}].$$

Proof. Condition (6.5b) can be written as $r < 1 < r(1 - Q)$; observe that $cd < 1$ entails $Q < 0$. From (6.25) we see that there exists $\varepsilon \in (0, X)$ such that $X_m, \ldots, X_{m+N-1} \leqslant \varepsilon$. By arguing as in the proof of Theorem 6.5 we obtain (6.10) once again where $f(\cdot)$ is given by (6.20). We now have $f(s) < s$ if $0 < s < X$ and Kuczma's theorem asserts that for such s, $f_k(s) \to 0$ as $k \to \infty$. Since $f'(0) = r$, Szekeres's theorem shows that the rate of convergence is given by $f_k(s) \sim \text{const.}\, r^k$, and the first assertions of the theorem follow. The second set of assertions follows in a similar way because $f(s) > s$ when $s > X$ and then $1 - f_k(s) \sim \text{const.}$ $(f'(1))^k$ and $f'(1) = 1/r(1 - Q) < 1$.

Theorem 6.13 does not eliminate the possible existence of an orbit for which $X_n - X$ changes sign along an infinite subsequence $\{n_i\}$ satisfying $n_{i+1} - n_i < N$. In all cases where a finite memory version of Model 2 has been fitted to experimental data we have found that the estimated parameters fall within the ambits of Theorems 6.10 or 6.11. We conjecture that when (6.5b) is satisfied then orbits satisfy the hypotheses of Theorem 6.13.

Assuming this conjecture to be true, the results established above allow us to draw a picture of the way in which the system behaves as c and d are varied but $d^* = A/B$ remains fixed. Our description will be given in terms of the maximum seed pool sizes x_n and y_n. The boundary equilibria now are $(0, L)$ and $(M, 0)$, and the point (X, Y) becomes (x_a, y_a) where

$$x_a = L\frac{c - 1/d^*}{c - 1/d} \quad \text{and} \quad y_a = M\frac{d - d^*}{d - c^{-1}}.$$

The experimental data has always given $d^* < 1$ and consequently we will assume this for the ensuing discussion. We partition the parameter set $\{c > 0, d > 0\}$ into six open regions together with their common boundaries; see Figure 6.2.

The hyperbola $cd = 1$ divides the parameter set into two regions so that in regions 2, 3 and 6 we have $cd > 1$. Now (6.5a) is equivalent to $c > 1/d^*$ and $d > d^*$ which together delineate region 3. Thus in region 3 the point (x_a, y_a) is globally attracting for all orbits $\{(x_n, y_n): x_1, y_1 > 0\}$. Similarly (6.5b) is equivalent to $c < 1/d^*$ and $d < d^*$ which delineate region 4 in which (x_a, y_a) is repelling, but both boundary equilibria have basins of attraction (or so we conjecture). Regions 1 and 2 correspond to the strict forms of (6.16) and (6.15), respectively, and here $(x_n, y_n) \to (0, M)$ geometrically fast. This convergence still obtains on the common boundary of regions 1 and 2, but the proof of Theorem 6.10 suggests a much slower rate of convergence.

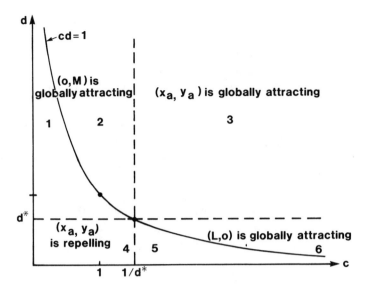

Figure 6.2. Nature of the equilibria as c and d vary.

Suppose now that d is fixed, $d > d^*$ and c increases from a small value. When $c < 1/d^*$ there is no possible internal point attractor and $(0, M)$ is a global attractor. This persists as the point (c, d) crosses from region 1 into region 2 but there is a qualitative change in the system dynamics because the curvature of the invariant curve $y = M\gamma(x/L)$ changes sign. When (c, d) approaches the boundary separating regions 2 and 3, $(0, M)$ continues to attract all orbits, but much more weakly as suggested by Theorem 6.10. As (c, d) enters region 3 the boundary equilibria are no longer attracting. The point global attractor (x_a, y_a) bifurcates from $(0, M)$ and x_a increases from 0 to L as c increases from $1/d^*$ to infinity. At the same time y_a decreases from M to zero.

In ending this section we remark that exact convergence rates for the strict inequality cases of Theorem 6.10 can be derived in a manner very similar to the proof of Theorem 6.7. For example, in the strict cases of (6.15) we can write

$$X_n = (1 + \delta_n)(c/B) \sum_{i=1}^{N} a_i X_{n-i}$$

if $n > N$, where now

$$\delta_n = \left[(1 - Q) \sum_{i=1}^{N} b_i X_{n-i} / (1 - Q X_{n-i}) - c \sum_{i=1}^{N} a_i X_{n-i} \right]$$
$$\div \sum_{i=1}^{N} (c a_i X_{n-i} + b_i Y_{n-i}).$$

Obviously $\delta_n \to 0$ and this allows the analysis of Theorem 6.7 to go through with only minor changes. In particular there exists $\eta > 1$ solving the polynomial equation $(c/B)\sum_{n=1}^{N} a_n \eta^n = 1$ and hence if the aperiodicity condition (6.12a) is satisfied then $X_n \sim$ const. η^{-n} as $n \to \infty$. Conditions (6.12c) and (6.12d) required for Theorem 6.7 are automatically satisfied here. Observe that $r^{-1/N} < \eta < r^{-1}$.

A similar analysis applies in the case (6.5b) to orbits which converge to the boundary equilibria. Thus orbits $\{(x_n, y_n)\}$ which are attracted to $(0, M)$ converge to it at a rate like η^{-n}. On the other hand, orbits attracted to $(L, 0)$ satisfy $y_n \sim$ const. $\tilde{\eta}^{-n}$ where $\tilde{\eta} > 0$ solves the equation $(d/A)\sum_{n=1}^{N} b_n z^n = 1$.

This leaves the case (6.5a) where we have an interior point attractor. The determination of the exact rate of convergence in this case involves a pair of discrete renewal equations, rather like the situation for Model 1A considered in Section 5.4. A simplified approximate analysis can be based on linearized versions of these equations. However, this analysis is not tied to the finite memory assumption and so we shall take this up in the next section where we allow $N = \infty$.

6.5   Local analysis of the infinite memory case

In this section we will perform an analysis of the linearized version of (6.20) when $N = \infty$. Even with the simplifications entailed by linearization we still need to supplement condition (6.5a), which ensured the existence of an internal attracting point for the finite memory case, with a strong condition on the softening rates. This is that they decay geometrically fast. We extend previously used notation by setting $A = \sum_{n \geqslant 1} a_n$ and $B = \sum_{n \geqslant 1} b_n$. The main result of this section follows.

> *Theorem 6.14.*
> *Suppose $N = \infty$ and A and B are finite.*
>
> (i) *If (6.5a) holds and there exists $\theta > 0$ such that $\sum_{n \geqslant 1} a_n (1 + \theta)^n$ and $\sum_{n \geqslant 1} b_n (1 + \theta)^n$ are finite, then (X, Y) is locally attracting and the boundary equilibria are locally repelling.*
>
> (ii) *If (6.5b) holds then (X, Y) is locally repelling and the boundary equilibria are locally attracting.*
>
> (iii) *If the strict version of (6.15) or (6.16) is satisfied (respectively, (6.17) or (6.18)), or if $cd = 1$ and $cA < B$ (respectively, $cd = 1$ and $B < cA$), then $(0, 1)$ (respectively, $(1, 0)$) is locally attracting and $(1, 0)$ (respectively, $(0, 1)$) is locally repelling.*

Proof. We will give details only for the case where (6.5) holds. Let $\delta_n = X_n - X$ and $\varepsilon_n = Y_n - Y$, and suppose that δ_n and ε_n are small compared to X and Y, respectively. We can write (6.1a) in the form

$$X_n = \frac{cAX + c \sum_{1 \leqslant i < n} a_i \varepsilon_{n-i} + o_1(1)}{cAX + BY + \sum_{1 \leqslant i < n} (ca_i \varepsilon_{n-i} + b_i \delta_{n-i}) + o_2(1)}$$

where

$$o_1(1) = -cX \breve{a}_n \quad \text{and} \quad o_2(1) = -cX \breve{a}_n - Y \breve{b}_n.$$

Setting $u_i = a_i/A$ and $v_i = b_i/B$ we can subtract X from each side of this equation and cancel the leading terms by using (6.7). Finally, if the second and higher order terms are omitted and we define

$$\rho = 1/r = B/cA,$$

there remains the linearized equation

$$\delta_n = \rho Y \sum_{i=1}^{n-1} u_i \delta_{n-i} - \rho X \sum_{i=1}^{n-1} v_i \varepsilon_{n-i} + \rho XY(\breve{v}_n - \breve{u}_n).$$

Recalling that $R = A/Bd$, we find that the corresponding equation for ε_n is

$$\varepsilon_n = -RY \sum_{i=1}^{n-1} u_i \delta_{n-i} + RX \sum_{i=1}^{n-1} v_i \varepsilon_{n-i} - RXY(\breve{v}_n - \breve{u}_n),$$

and each holds for $n \geqslant 2$.

This pair of equations can be solved by introducing the generating functions

$$\Delta(t) = \sum_{n \geqslant 0} \delta_{n+1} t^n, \quad E(t) = \sum_{n \geqslant 0} \varepsilon_{n+1} t^n,$$

$$U(t) = \sum_{n \geqslant 1} u_n t^n \quad \text{and} \quad V(t) = \sum_{n \geqslant 1} v_n t^n.$$

Clearly,

$$\Delta(t) - \delta_1 = \rho Y \Delta(t) U(t) - \rho X E(t) V(t) + \rho XY(V(t) - U(t))/(1 - t)$$

and there is a similar equation for $E(t)$. These equations can be expressed as

$$[1 - \rho YU(t)]\Delta(t) + \rho XV(t)E(t) = P(t)$$

and

$$RYU(t)\Delta(t) + [1 - RXV(t)]E(t) = Q(t) \tag{6.26}$$

where

$$P(t) = \delta_1 + \rho XY(V(t) - U(t))/(1 - t)$$

and

$$Q(t) = \varepsilon_1 - RXY(V(t) - U(t))/(1 - t).$$

The determinant of this system is

$$D(t) = 1 - \rho Y U(t) - R X V(t).$$

Clearly, $D(\cdot)$ is decreasing, $D(0) = 1$, and since

$$X = (1 - \rho)/(1 - \rho R) \quad \text{and} \quad Y = (1 - R)/(1 - \rho R),$$

we have

$$D(1) = (1 - \rho)(1 - R)/(1 - \rho R).$$

In particular $D(1) = (1 - \rho)Y$ and since (6.5a) implies that $\rho > 1$, it follows that $D(t)$ is positive in $[0, 1]$. Similarly if (6.5b) holds then there exists $q \in (0, 1)$ such that $D(q) = 0$.

When the condition under (i) is satisfied, $D(t)$ and $(V(t) - U(t))/(1 - t)$ are defined and finite if $0 \leqslant |t| < 1 + \theta$. Consequently, by choosing θ small enough we ensure that $D(t) > 0$ if $1 \leqslant t \leqslant 1 + \theta$. It follows that the system (6.25) has solutions $\Delta(t)$ and $E(t)$ which are holomorphic in the disc $0 \leqslant |t| < 1 + \theta$. It follows that δ_n, $\varepsilon_n \to 0$ ($n \to \infty$) geometrically fast and hence (X, Y) is locally attracting.

If (6.5b) holds then similar considerations show that (6.25) has a pair of solutions which are holomorphic only for $|t| < q$. In particular δ_n and ε_n increase like q^{-n} and it follows that (X, Y) is locally repelling.

Now consider behaviour near $(0, 1)$. If $(\tilde{X}_n)$ denotes the linearization of (X_n) we have, from (6.1a), $\tilde{X}_n = r \sum_{i=1}^{n-1} u_i \tilde{X}_{n-i}$, where $\tilde{X}_1 = X_1$, whence

$$X(t) = \sum_{n \geqslant 0} \tilde{X}_{n+1} t^n = X_1/(1 - r U(t)).$$

It follows that for each n, $\tilde{X}_n$ has the same sign as X_1. The hypotheses imply that $r < 1$ and $\sum \tilde{X}_{n+1} < \infty$. In particular $\tilde{X}_n \to 0$. By setting $\varepsilon_n = Y_n - 1$, it is easily shown that the linearized version of (6.1b) is $\varepsilon_n = -(R/r) \sum_{i=1}^{n-1} u_i \tilde{X}_{n-i} = -\tilde{X}_n/cd$, and it follows that $(0, 1)$ is locally attracting. The other cases are handled similarly.

The above analysis gives an indication of rates of convergence to attracting points. Under the conditions of (i) above, for example, when θ can be chosen large enough to ensure the existence of a number $\eta > 1$ such that $D(\eta) = 0$, then δ_n and ε_n will be asymptotically proportional to η^{-n} as $n \to \infty$. Similarly, the above linearized analysis about $(0, 1)$ can be extended to show that $\tilde{X}_n \to 0$ geometrically fast provided there exists $\theta > 0$ such that $1/r \leqslant U(1 + \theta) < \infty$.

Apart from Model 1A, models for which $N = \infty$ are not likely to have biological significance. Infinite memory models for which softening rates decay so slowly that A or B are no longer finite are even less likely to be biologically relevant. From the mathematician's point of view, however, it is interesting to ask whether the results and insights provided by Theorem 6.14 persist when A and/or B are infinite. We do not intend giving a detailed examination of these cases, rather, we simply look at a couple of particular instances which can be regarded as boundary cases of Model 1A. The smooth behavior seen above persists in that orbits converge to certain points of the state-space, but rates of convergence are slower.

We consider first the case where softening rates for Strain 1 decrease geometrically fast, $a_i = \alpha^i$ where $0 < \alpha < 1$, but for Strain 2 they are constant; $b_i = \beta$. If we set $T_n = \sum_{i=1}^{n} \alpha^i X_{n+1-i}$ and $U_n = \hat{Y}_n (n \geqslant 1)$ then $\alpha X_n = T_n - \alpha T_{n-1}$ and (6.1) can be recast as: for $n \geqslant 2$

$$T_n = \alpha T_{n-1}(1 + c/(c\alpha T_{n-1} + \beta U_{n-1})) \qquad (6.27a)$$

and

$$U_n = U_{n-1}(1 + \beta d/(T_{n-1} + \beta d U_{n-1})). \qquad (6.27b)$$

This system is very similar to Model 1A as formulated by (3.1) and it is easily analysed as follows.

Since $0 < X_n, Y_n < 1$ $(n \geqslant 2)$, the sequence $\{T_n\}$ is bounded and $\{U_n\}$ is strictly increasing. If $U_n \to U < \infty$, it follows from (6.27b) that $T_{n-1} \to T$, say, whence $1 = 1 + \beta d/(T + \beta d U)$, a contradiction. Thus $U_n \to \infty$ and it follows from (6.27a) that $T_n/T_{n-1} \to \alpha$. Consequently, for any $\bar{\alpha} \in (\alpha, 1)$, $T_n = O(\bar{\alpha}^n)$. In particular, $X_n \to 0$, whence Theorem 6.2 yields $Y_n \to 1$. This implies $U_{n-1} \sim n$ and consequently we can express (6.27a) as

$$T_n = \alpha T_{n-1}(1 + c/(\beta n + \varepsilon_n))$$

where $\varepsilon_n = o(n)$. Iteration yields

$$T_n = T_1 \alpha^{n-1} \prod_{j=2}^{n} (1 + c/(\beta j + \varepsilon_j))$$

and it is easily shown that the product is asymptotically proportional to $n^{c/\beta}$ as $n \to \infty$. It follows that $T_n \sim \text{const.} \, \alpha^n n^{c/\beta}$. We conclude, then, that

$$X_n \sim cT_{n-1}/\beta n, \quad 1 - Y_n \sim T_{n-1}/\beta dn$$

and $(1 - Y_n)/X_n \to cd$. These results can be compared with Theorem 4.2. Observe the presence of the algebraic factor $n^{c/\beta}$ which does not arise in Model 1A.

Next, we consider the case where both strains have constant softening rates, denoted by α and β for Strains 1 and 2, respectively. If we set $q = \alpha c/\beta$ and consider the de Wit case then Theorem 6.2, which is still valid, allows us to write (6.1a) as a univariate difference equation,

$$X_n = q\hat{X}_{n-1}/(n - (1 - q)\hat{X}_{n-1} + X_1 + Y_1 - 2). \qquad (6.28)$$

Let $q < 1$ and set $\varepsilon = qX_1/(qX_1 + Y_1)$. Then $X_2 = \varepsilon$. Now if $X_2 > \ldots > X_{n-1} > X_n$ then $\hat{X}_n < \varepsilon + \hat{X}_{n-1}$ and since the right-hand side of (6.28) is an increasing function of $\hat{X}_{n-1}$ we obtain

$$X_{n+1} < \frac{q(\hat{X}_{n-1} + \varepsilon)}{n + 1 - (1 - q)\hat{X}_{n-1} + X_1 + Y_1 - 2 - (1 - q)\varepsilon} < X_n.$$

A similar calculation shows that $X_3 < \varepsilon$ and hence $\{X_n : n \geqslant 2\}$ is a decreasing sequence. It follows that $\{X_n\}$ converges to a limit which, by the following reasoning must be zero. If the limit $x > 0$, it follows from (6.27) that $x = 1$ and this is impossible since $X_n < 1$ and $\{X_n : n \geqslant 2\}$ is decreasing. Theorem 6.2 now yields $Y_n \to 1$.

To obtain the rate of convergence we observe that $\hat{X}_{n-1}/n \to 0$ and hence (6.28) implies $X_n \sim q\hat{X}_n/n$. It can be deduced from this relation that for j' large enough,

$$\log \hat{X}_n = \sum_{j=j'}^{n} (X_j/\hat{X}_j) + O(1),$$

and hence $X_n \sim \text{const.} \, n^{q-1}$. Consequently we obtain behavior similar to that of Model 1A in the de Wit case but here the boundary equilibrium is approached at an algebraic rate, in contrast to the geometric rate we found for Model 1A.

If $q > 1$ it is clear that orbits approach $(1, 0)$ algebraically fast. When $q = 1$, Equation (6.8) can be solved explicitly for X_n and it can then be shown that $X_n \to KX_1$ where

$$K = \prod_{j=0}^{\infty} \left(1 - \frac{X_1 + Y_1 - 1}{(j + X_1 + Y_1)(j + 2)} \right).$$

This is similar to Model 1A with co-incident isoclines, but here $X_n - KX_1 \sim \text{const.} \, n^{-1}$, in contrast to the geometric rate for Model 1A – see Theorem 5.7.

When $cd \neq 1$, we need to examine several cases which are analogous to those in Theorem 6.4. However, having made our point with the above examples, we will spare the reader a detailed analysis of these additional cases.

7

Application of the models

7.1 Introduction

In practice, we are interested in three main aspects of a strain mixture: (i) Does one strain ultimately win, if so which, or is there a (stable or unstable) equilibrium? (ii) How long does it take the losing strain to disappear, or how fast is the approach to equilibrium?. (iii) How do the strains interact in the short term? The answers to questions (i) and (ii) are contained in Chapters 3–6, and they can be applied once estimates of the parameters used in the models have been obtained. In Section 2 of this chapter we explain how these "standard values" were obtained, and tabulate them. Section 3 summarises the analytical results of the previous chapters and shows how they can be used to predict long-term behavior of our standard mixtures.

The remainder of this chapter is devoted to various aspects of the short-term behavior of a mixture. In Section 5 the various models are compared for the standard values of the parameters. It turns out that, for our data, the models are little different, and Model 1A is an adequate description of their behavior. Nevertheless the models can potentially differ greatly in their short-term behavior, as is demonstrated in Section 5, and this may be important in other applications, if not for our standard mixtures. The main tool used in Section 5 is a collection of computer programs in which the values of new and total seed production, established plants, and so forth, are calculated recursively for given values of the parameters.

Some of the results of this chapter have important implications in practise, apart from the obvious questions relating to dominance of a strain, because we can investigate the roles played by the different parameters of the models in establishing dominance. We reveal some effects that were previously unknown; for example, the summer survival parameters P_S and Q_S turn out to be unexpectedly important. Agronomists previously thought that seed softening characteristics had most

131

impact, and breeding programmes in Western Australia reflected this belief, in that strains were being selected for their hard-seededness. We show in Section 4, where the effect on dominance of the interaction between summer survival and seed softening characteristics is investigated, that it may be pointless in some cases to try to improve the competitiveness of a strain by changing its rate of seed softening.

7.2   The standard data

In Rossiter et al. (1985) data were collected from a number of sources to estimate the parameters required to apply Model 1A to 6 binary mixtures. For the three "main mixtures", Dwalganup–Daliak, Yarloop–Seaton Park, and Seaton Park–Midland B, a reliable set of data was assembled from various experiments: (i) monoculture yields were determined from Taylor et al. (1984); (ii) de Wit constants k_{12} and k_{21} were obtained from a series of competition experiments discussed by Rossiter and Palmer (1981); (iii) estimates of summer survival, establishment and softening rates were obtained by averaging over six years the values obtained from an experiment in which the three mixtures were observed from a 50/50 mixture sown in 1970, and grazed under field conditions till 1977. These estimates were judiciously combined with other information available on them from the long experience of Western Australian agronomists with these strains. These values, given in Table 7.1, thus represent the distillation of many years of scientific experience.

For another three mixtures, Dwalganup–Northam A, Daliak–Dinninup, and Geraldton–Dinninup, parameter estimates of somewhat lesser reliability were also available for Model 1A, as described in Rossiter et al. (1985). These are also given in Table 7.1. In Table 7.2 we have assembled estimates of the parameters required to apply Models 1 and 2 to the three main mixtures. Some of this information, such as the age-dependent rates of softening, was gleaned from Taylor et al. (1984) and Taylor (personal communication) and some, such as the rates of transition of burr seed to free, is little more than an educated guess. (We thank Dr R. C. Rossiter for supplying these.) In principle, of course, all these parameters could be measured. We also give, in Table 7.3, some of the derived parameters for Models 1 and 2 which show how softening rates vary with time; see Chapters 2 and 3 for the notation. The numbers in Table 7.3 are rounded from values calculated using double precision arithmetic, and these more accurate values were used for calculating the entries of Table 7.5.

Table 7.1. *Parameters for Model 1A*

	Dwalganup–Daliak		Yarloop–Seaton Park	
P_S Q_S	.82	.87	.75	.82
P_E Q_E	.35	.47	.44	.44
P_A Q_A	.70	.70	.75	.70
$P_V{}^a$	.30	.30	.30	.30
W	1.18	1	1.38	1
$P(F\mid B)$ $Q(F\mid B)^b$	.25	.25	.15	.15
$P(S\mid B)$ $Q(S\mid B)$	.36	.36	.48	.48
$P(S\mid F)$ $Q(S\mid F)$	.51	.51	.63	.63
M_1 M_2	194.00	291.00	104.00	144.00
k_{12} k_{21}	1	1	1	1

	Seaton Park–Midland B		Dwalganup–Northam A	
	.82	.87	.82	.82
	.44	.44	.35	.35
	.70	.70	.70	.70
	.30	.30	.30	.30
	1.37	1	1	1
	.15	.15	.25	.25
	.48	.35	.36	.30
	.63	.50	.51	.45
	144.00	197.00	194.00	194.00
	1	1	.594	1/.594

	Daliak–Dinninup		Geraldton–Dinninup	
	.87	.87	.87	.87
	.47	.47	.44	.44
	.70	.70	.70	.70
	.30	.30	.30	.30
	1	1	1	1
	.25	.25	.25	.25
	.36	.40	.33	.40
	.51	.55	.48	.55
	291.00	155.00	250.00	155.00
	.3	1/.3	.25	1/.25

[a] Value given for eighth and subsequent years. For years 1–7, $P_V = 1, 1, 1, 1, .9,$ .75, .50.
[b] Values given for fifth and subsequent years.

7.3 Applications of the models

It is convenient to summarise here the analysis of Model 1A. This depends on the quantities k_{12}/ρ and $k_{21}\rho$, and the only possible outcomes are:

(A) $k_{12}/\rho \geqslant 1$ and $k_{21}\rho \leqslant 1$, and then Strain 1 wins; or
(B) $k_{12}/\rho \leqslant 1$ and $k_{21}\rho \geqslant 1$, and then Strain 2 wins; or

Table 7.2. *Parameters for Models 1 and 2*

	Dwalganup–Daliak		Yarloop–S. Park		S. Park–Midland B	
$P_S \quad Q_S$	.82	.87	.75	.82	.82	.87
$P_E \quad Q_E$	.35	.47	.44	.44	.44	.44
$P_A \quad Q_A$	.70	.70	.75	.70	.70	.70
$P_V{}^a$	.30	.30	.30	.30	.30	.30
W	1.18	1.00	1.38	1.00	1.37	1.00
For Models 1 and 1A:						
$P(F\mid B),\ Q(F\mid B)^b$	.25	.25	.15	.15	.15	.15
For Model 2:						
$P_1(F\mid B),\ Q_1(F\mid B)$	.00	.00	.00	.00	.00	.00
$P_2(F\mid B),\ Q_2(F\mid B)$	.10	.10	.10	.10	.10	.10
$P_3(F\mid B),\ Q_3(F\mid B)$	.20	.20	.20	.20	.20	.20
$P_4(F\mid B),\ Q_4(F\mid B)$	.40	.40	.40	.40	.40	.40
$P_5(F\mid B),\ Q_5(F\mid B)$	.70	.70	.70	.70	.70	.70
$P_6(F\mid B),\ Q_6(F\mid B)$	1.00	1.00	1.00	1.00	1.00	1.00
For Models 1, 2 but not 1A						
$P_1(S\mid B),\ Q_1(S\mid B)$	.25	.25	.40	.40	.40	.20
$P_2(S\mid B),\ Q_2(S\mid B)$	.30	.30	.50	.50	.50	.30
$P_3(S\mid B),\ Q_3(S\mid B)$	.40	.40	.60	.60	.60	.40
$P_4(S\mid B),\ Q_4(S\mid B)$	.60	.60	.80	.80	.80	.60
$P_5(S\mid B),\ Q_5(S\mid B)$	.80	.80	1.00	1.00	1.00	.80
$P_6(S\mid B),\ Q_6(S\mid B)$	1.00	1.00	–	–	–	1.00
$P_1(S\mid F),\ Q_1(S\mid F)$	.40	.40	.55	.55	.55	.35
$P_2(S\mid F),\ Q_2(S\mid F)$	.45	.45	.65	.65	.65	.45
$P_3(S\mid F),\ Q_3(S\mid F)$	.55	.55	.75	.75	.75	.55
$P_4(S\mid F),\ Q_4(S\mid F)$	.75	.75	.95	.95	.95	.75
$P_5(S\mid F),\ Q_5(S\mid F)$	.95	.95	1.00	1.00	1.00	.95
$P_6(S\mid F),\ Q_6(S\mid F)$	1.00	1.00	–	–	–	1.00
$M_1 \quad M_2$	194.00	291.00	104.00	144.00	144.00	197.00
$k_{12} \quad k_{21}$	1.00	1.00	1.00	1.00	1.00	1.00

[a] Value given for eighth and subsequent years. For years one–seven, $P_V = 1,\ 1,\ 1,\ 1,\ .9,\ .75,\ .5$.
[b] Value given for fifth and subsequent years.

(C) $k_{12}/\rho \geqslant 1$ and $k_{21}\rho \geqslant 1$, and there is a stable interior equilibrium; or

(D) $k_{12}/\rho \leqslant 1$ and $k_{21}\rho \leqslant 1$, and there is an unstable interior equilibrium.

We tacitly assume that at least one of the inequalities above is strict; there is no likelihood of observing the exact equality $k_{12}/\rho = 1$. In practise, however, we sometimes have situations which are very close to exact equalities, see for example, our discussion of Seaton Park–Midland B below.

Table 7.3. *Derived parameters for Models 1 and 2*

	Model 1				Model 2			
	S_i	T_i	a_i	b_i	S_i	T_i	a_i	b_i
Dwalganup–Daliak	.2357	.2501	13.222	23.947	.2050	.2175	11.498	20.823
$L = M_1 P_V = 58.2\ (n \geqslant 8)$	.1617	.1820	9.069	17.426	.1589	.1788	8.909	17.120
$M = M_2 P_V = 87.3\ (n \geqslant 8)$	.1139	.1360	6.386	13.020	.1245	.1487	6.985	14.239
$P_E P_A W =\ $ .2891	.0765	.0970	4.292	9.284	.0884	.1120	4.957	10.723
$Q_E Q_A =\ $ .3290	.0299	.0402	1.677	3.847	.0316	.0425	1.774	4.071
$N =\ $ 6	.0048	.0068	.267	.649	.0021	.0031	.121	.293
$A = \Sigma a_i,\ B = \Sigma b_i$			34.912	68.172			34.243	67.270
Yarloop–Seaton Park	.3169	.3464	15.008	15.366	.3000	.3280	14.208	14.547
$L = M_1 P_V = 31.2\ (n \geqslant 8)$	.1697	.2029	8.039	8.999	.1738	.2078	8.232	9.215
$M = M_2 P_V = 43.2\ (n \geqslant 8)$	.0724	.0946	3.430	4.198	.0784	.1025	3.713	4.545
$P_E P_A W =\ $ .4554	.0271	.0387	1.283	1.717	.0291	.0416	1.380	1.847
$Q_E Q_A =\ $ .3080	.0044	.0069	.208	.304	.0031	.0048	.146	.214
$N =\ $ 5								
$A = \Sigma a_i,\ B = \Sigma b_i$			27.967	30.583			27.681	30.369
Seaton Park–Midland B	.3464	.1936	21.051	11.745	.3280	.1740	19.930	10.558
$L = M_1 P_V = 43.2\ (n \geqslant 8)$	.2029	.1898	12.328	11.516	.2078	.1907	12.627	11.573
$M = M_2 P_V = 59.1\ (n \geqslant 8)$	.0946	.1466	5.751	8.892	.1025	.1586	6.226	9.626
$P_E P_A W =\ $ .4220	.0387	.1085	2.352	6.583	.0416	.1195	2.531	7.249
$Q_E Q_A =\ $ .3080	.0068	.0471	.416	2.856	.0048	.0454	.294	2.752
$N =\ $ 6	.000	.0088	.000	.536	.0000	.0030	.000	.183
$A = \Sigma a_i,\ B = \Sigma b_i$			41.899	42.128			41.605	41.941

Now if $k_{12}k_{21} = 1$, only (A) or (B) can occur, so one strain must win in this situation, except for the case (having no likelihood in practise) where $k_{12} = \rho$. Case (C) can only occur if $k_{12}k_{21} > 1$, that is, one strain is more "competitive" than the other in the de Wit part of the model. Case (D) only occurs if $k_{12}k_{21} < 1$.

Let us explain how we deduced (A)–(D) from the results of the previous chapters. One of the key conclusions of Chapters 3 and 4 is that the final outcome can be predicted from the configuration of the isoclines $\mathscr{I}_F$ and $\mathscr{I}_G$, which are described by the following equations:

$$\mathscr{I}_F: \quad k_{12}p + q = k_{12};$$
$$\mathscr{I}_G: \quad p + k_{21}q = \rho k_{21}. \tag{7.1}$$

For each of the main mixtures we have $k_{12} = k_{21} = 1$, and then $\mathscr{I}_F$ and $\mathscr{I}_G$ are the parallel lines $p + q = 1$ and $p + q = \rho$, respectvely. If (A) or (B) holds then the isocline $\mathscr{I}_F$ lies above $\mathscr{I}_G$, or $\mathscr{I}_G$ lies above $\mathscr{I}_F$ (see (4.8)) so Theorem 4.1 tells us that (1, 0) or (0, ρ) is GAS for (p_n, q_n), that is, one of the strains wins from any starting position. If neither (A) nor (B) holds then Chapter 4 does not apply and we turn to Chapter 5.

When (C) holds, Theorem (5.2) (see (5.4) as well) tells us that

$$(\bar{p}, \bar{q}) = \left(\frac{k_{21}(k_{12} - \rho)}{k_{12}k_{21} - 1}, \frac{k_{12}(\rho k_{21} - 1)}{k_{12}k_{21} - 1}\right) \tag{7.2}$$

is a GAS interior equilibrium for (p_n, q_n). Finally if (D) holds, we are in the domain of Section 6 in Chapter 5, and we have an unstable interior equilibrium $(\bar{p}, \bar{q})$. In cases (C) and (D) this equilibrium point is obtained by simultaneous solution of the isocline equations (7.1).

To translate back to the original system, we have for the maximum seed pools (t_n, u_n), that (see (3.4))

$$t_n = \left(\frac{L}{1 - \alpha}\right)p_n, \quad u_n = \left(\frac{U}{1 - \alpha}\right)(q_n/Y)$$

where the various symbols are defined in Section 3.1. Thus in Outcomes (A)–(C) we have $(t_n, u_n) \rightarrow (t_a, u_a)$ where in Outcome

(A) $t_a = L/(1 - \alpha)$ and $u_a = 0$;
(B) $t_a = 0$ and $u_a = M/(1 - \beta)$;

and for (C) the long term values are

$$t_a = L\bar{p}/(1 - \alpha) \quad \text{and} \quad u_a = U\bar{q}/Y(1 - \alpha) = M\bar{q}/\rho(1 - \beta),$$

$$\tag{7.3}$$

where we have used (3.5) to obtain the last equality.

For Outcomes (A) or (B) we have rates of convergence from Theorem 4.2. Let us assume that Outcome B holds, that is Strain 2 wins, which is the case for our standard data. Theorem 4.2 then tells us that $\mathscr{L} = \lim_{n \to \infty} (\rho - q_n)/p_n$ exists. In addition, we compare β with $\Delta = \alpha + Uk_{12}(1 - \beta)/V$ to determine the value of $\mathscr{L}$. For our standard data it transpires that $\beta < \Delta$, which is case (a) in Theorem 4.2, and hence Strain 1 vanishes at a geometrically fast rate (see (4.18)):

$$p_n \sim p_1 \Gamma \Delta^n$$

and Strain 2 eventually increases up to its limit at the same rate, that is,

$$\rho - q_n \sim p_1 \Gamma \mathscr{L} \Delta^n.$$

Thus the rates of approach of t_n and u_n to their equilibria are also at the rate of Δ^n. The rate of approach under (C) to the stable interior equilibrium is given by Theorem 5.5.

Case (D), of an unstable equilibrium is interesting, although it does not occur with our standard mixtures. A possible scenario for its application is one in which Strain 2 is more competitive than Strain 1, but Strain 1 exerts a toxic effect on Strain 2. Thus, with a large enough

Table 7.4. *Stability analysis for Model 1A*

	k_{12}/ρ	$k_{21}\rho$	Outcome	u_a	y_a	β	Δ
Dwal.–Daliak	.519	1.926	B	183.5	87.3	.524	.757
Yar.–S. Park	.922	1.084	B	73.0	43.2	.408	.951
S. Park–Mid. B	.977	1.024	B	130.1	59.1	.546	.986
Dwal.–Nort. A	.632	1.583	B	127.4	58.2	.543	.814
Daliak–Dinn.	.549	1.821	B	91.1	46.5	.489	.786
Gerald.–Dinn.	.385	2.600	B	91.1	46.5	.489	.723
				(t_a, u_a)		(x_a, y_a)	R
S. Park–Mid. B[a]	1.045	1.024	C (47.9, 46.7)			(28.4, 21.7)	.992

Dwal. = Dwalganup, Yar. = Yarloop, S. Park = Seaton Park, Mid. B = Midland B, Nort. A = Northam A, and Gerald. = Geraldton.
[a] $k_{12} = 1.07$, $k_{21} = 1.00$, and all other parameter values are standard data values for Seaton Park–Midland B.

component of Strain 1 initially in the mixture, Strain 1 could win despite the competitive superiority of Strain 2.

The first six lines of Table 7.4 give the derived parameter values required to characterise the long term behavior of the six strain mixtures under Model 1A. In each case Outcome (B) holds, that is, Strain 2 eliminates Strain 1. In addition $\beta < \Delta$ for each, so the losing strain disappears at a rate proportional to Δ^n, and the winning strain approaches its limit t_a (also given in Table 7.4) at exactly the same rate. Note that for the first listed Seaton Park–Midland B mixture, $\Delta = 0.986$, very close to 1, and in fact it takes up to 600 years for Seaton Park to disappear for the standard data values.

We note that for this mixture k_{12}/ρ and $k_{21}\rho$ are very close to unity, and a slight change in a value, for example, of k_{12} to 1.07 but leaving the other parameters unchanged, results in Outcome (C) for this mixture; see line 7 of Table 7.4. That is, there is now a stable equilibrium in which the two strains coexist indefinitely. Thus the standard values for this strain pair are very close to the boundary separating Outcomes (B) and (C). Notice that the long-term maximum seed pool of Midland B is substantially reduced when k_{12} increases from unity to 1.07. The entry in the Δ column of this line is the larger eigenvalue, denoted here by R, of the Jacobian matrix (3.14) of Model 1A evaluated at the interior equilibrium. As shown in Theorem 5.5 this quantity determines the speed of convergence to the interior equilibrium. We note that this occurs even more slowly than for the case where $k_{12} = 1$.

Let x_a and y_a denote the long term new seed productions of the competing strains. It follows from (2.14) that, in the notation of Section

3.1, $x_a = (1 - \alpha)t_a$ and $y_a = (1 - \beta)u_a$. Thus we have for Outcomes:

(A) $x_a = L$ and $y_a = 0$;
(B) $x_a = 0$ and $y_a = M$

and for Outcome (C) we have, from (7.2) and (7.3),

$$x_a = L\frac{k_{21}(k_{12} - \rho)}{k_{12}k_{21} - 1} \quad \text{and} \quad y_a = M\frac{k_{12}(k_{21} - 1/\rho)}{k_{12}k_{21} - 1}. \tag{7.4}$$

In Table 7.4 we have also listed the long-term new seed production for the winning strain when Outcome (B) occurs, and for each strain when Outcome (C) occurs. Below we will compare these values with those predicted by Models 1 and 2 for the main mixtures.

In Figures 7.1–7.7 we have illustrated some of these results for the Dwalganup–Daliak and Seaton Park–Midland B main mixtures. We used the PC package Phaser (Koçak (1986)) to plot phase portraits and graphs of p_n and q_n for a variety of starting values and time intervals. The quantities p_n and q_n were used instead of t_n and u_n because the former choice gives the most efficiently parametrised model and the package seemed to be much happier with this situation. In addition the results so obtained can be compared directly with the theoretical predictions of Chapters 4 and 5. Plots for (t_n, u_n) will differ only in scale from those given here. The reader should remember that our calculations are carried out with P_V set at its limiting value of 0.3, that is, we have ignored the initial transient variation of this quantity. Of course, this makes no difference to the long-term outcome; we will comment below on changes to the short term picture.

Figures 7.1 and 7.3 show phase portraits $[(p_n, q_n)$ versus $n]$ for the above-mentioned mixtures with initial positions around the edge of the portraits, as shown, except that starting values near the axes were set at 0.1, for example, $(p_1, q_1) = (1, 0.1)$. Figures 7.2 and 7.4 show plots of p_n and q_n against n. The results for Dwalganup–Daliak are unremarkable since, in qualitative form, they are exactly as would be expected from the discussions in Chapter 4. For all starting points shown the long-term outcome is essentially attained after twenty years.

If P_V is held constant at 0.3 then the maximum seed pool numbers t_n and u_n show the same qualitative behavior as illustrated in the figures. Our field data indicate that P_V decreases from unity to 0.3 over eight years and from Equations (3.1) it is evident that this has the effect of slowing down initial decreases from relatively large values of t_1 or u_1, or of speeding up initial increases from small values. This observation applies to any of the mixtures.

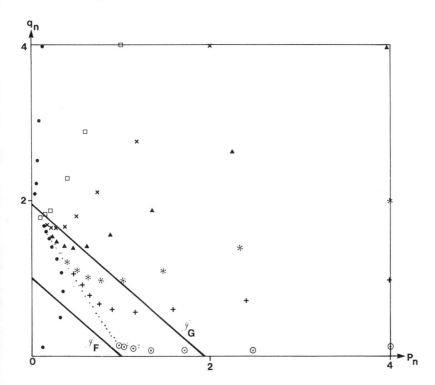

Figure 7.1. Phase portraits for the Dwalganup–Daliak mixture with $n = 1$–20, $\alpha = 0.49405$, $\beta = 0.5242$ and $\rho = 1.92587$.

We know from previous discussion that the Seaton Park–Midland B main mixture has isoclines which are nearly coincident [because ρ is close to unity] and that this mixture approaches its long term outcome very slowly indeed. Figures 7.3 and 7.4 show that this is attained in the order of hundreds of years. Actual times differ quite markedly with initial conditions. For example, the boundary equilibrium is substantially attained by 300 years when the system starts from $(0.1, 0.1)$, by 550 years when starting from $(4, 0.1)$, and in only 50 years when it starts from $(0.1, 4)$. The differences depend on the points at which the trajectories hit the strip between the isoclines. Since motion through this strip toward the equilibrium is very slow, orbits entering the strip near the p-axis take far longer to approach the equilibrium than do orbits which enter the strip nearer the boundary equilibrium.

This mixture exhibits another unexpected feature. Figures 7.3 and 7.4 show that the orbits reach the region between the isoclines in only a few years and then abruptly change their direction, thereafter moving slowly toward the attracting boundary equilibrium. On the scale of Figure 7.4 the turning points seem to occur about the same time,

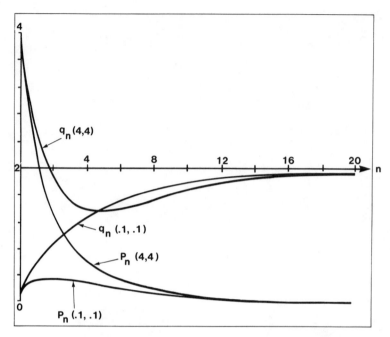

Figure 7.2. p_n and q_n against n for Dwalganup–Daliak with the indicated initial conditions and parameter values in Figure 7.1.

around year seven or eight. Dilation of the time scale shows that when the initial point is $(0.1, 0.1)$, the sequence (p_n) is maximised at $n = 9$, and (q_n) is minimised at $n = 11$ when the initial point is $(1, 1)$.

Figures 7.5, 7.6, and 7.8 illustrate behavior of the Seaton Park–Midland B mixture with k_{12} increased to 1.07. As we know, there now is an interior GAS equilibrium point $(\bar{p}, \bar{q})$, and

$$\bar{p} = .65896 \quad \text{and} \quad \bar{q} = .36502.$$

The phase portrait, Figure 7.5, is drawn with maximum coordinate values $p = q = 1$ so as to provide good resolution in a neighborhood of the equilibrium. We see again that orbits enter the wedge shaped regions between the isoclines in only a few years. Thereafter they converge slowly to the interior equilibrium, apparently showing transverse fluctuations within an even more narrow band inside the wedges. Limited screen resolution makes it quite difficult to determine the detailed structure of this band. The theory of Chapter 5 shows that orbits approach $(\bar{p}, \bar{q})$ by gliding along a straight line which lies within the wedges. A reasonable conjecture is that high magnification of a small region around $(\bar{p}, \bar{q})$ should reveal converging orbits forming a fibred structure. However, Figure 7.6 suggests that converging orbits

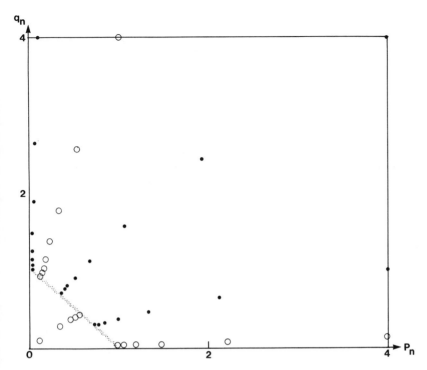

Figure 7.3. Phase portraits for the Seaton Park–Midland B mixture with $n = 1$–300, $\alpha = 0.40795$, $\beta = 0.54590$ and $\rho = 1.02388$.

quickly come very close to the asymptotic line and move toward the equilibrium almost on this line. In this figure we have indicated the various approach times to the equilibrium as functions of the starting positions.

By way of contrast, Figure 7.7 shows the phase portrait for the Dwalganup–Daliak mixture with $k_{12} = 3$ which gives a GAS internal equilibrium at $(0.53707, 1.38881)$. For this case the orbital configurations agree with those expected from the theoretical predictions of Chapter 5. These predictions demonstrate that there is no qualitative difference between the behaviors illustrated in Figures 7.5 and 7.7. The apparent atypical behavior of the Seaton Park–Midland B mixture arises solely from the near coincidence of its isoclines.

Figure 7.8 shows plots of p_n and q_n as functions of n for three pairs of initial conditions. The initial positions $(0.1, 0.1)$ and $(1, 1)$ manifest the short-duration and rapidly changing transient phase discussed above. The initial position $(0.1, 1)$ is very close to the region between the isoclines, and this is reached in only a few steps with a subsequent slow approach to the internal equilibrium.

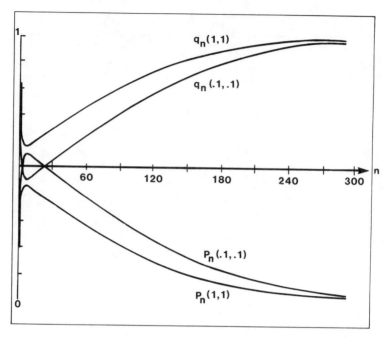

Figure 7.4. p_n and q_n against n for Seaton Park–Midland B with the indicated initial positions and parameter values in Figure 7.3.

We now consider Models 1 and 2, and in the notation of Chapter 6 define

$$\rho = B/A. \tag{7.5}$$

In the following discussion we will only consider the finite memory case, except where the softening rates S_i and T_i are such as to admit Model 1A. In this case Theorems 6.5, 6.10 and 6.11 together imply that *Outcomes (A)–(C) at the beginning of this section continue to apply to Models 1 and 2.* Remember that we tacitly assume at least one strict inequality in each of these assertions. To see that (A)–(C) really do follow from these theorems we restrict our considerations to Outcome (B) and argue as follows. If $k_{12}k_{21} > 1$ then our assumptions imply that (6.15) is satisfied, but if $k_{12}k_{21} < 1$, then (6.16) holds. In either case Theorem 6.10 assures us that Strain 2 wins. If $k_{12}k_{21} = 1$ then our tacit assumption of at least one strict inequality in Outcome (B) implies that both are strict and then Case (i) of Theorem 6.5 tells us that Strain 2 wins.

As mentioned in Chapter 6 it is not clear that Outcome (D), as a global statement, carries over to Models 1 and 2. Theorem 6.14 (ii) says that the conditions in Outcome (D) imply that the interior equilibrium

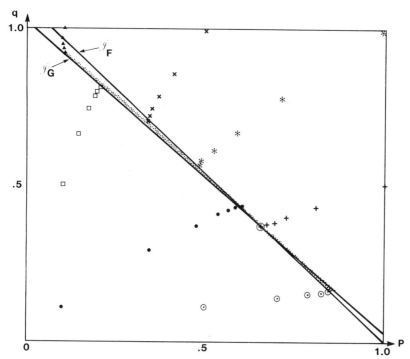

Figure 7.5. As for Figure 7.3 but with $k_{12} = 1.07$. ○ indicates the interior equilibrium.

is locally repelling. Remember that, strictly speaking, Models 1 and 2 do not in general have interior equilibria. In this chapter, however, we will relax our use of language to the extent of calling the point defined by (6.7) an interior equilibrium. None of the standard data falls within the ambit of Outcome (D), and our limited experience with synthetic parameter values suggests that under these conditions orbits behave according to Theorems 6.12 or 6.13, or a version of the latter, and one strain will win at the expense of the other, with the winner being determined by the initial proportions of the mixture.

We see also that all models predict exactly the same expressions for x_a and y_a whenever Outcomes (A), (B) or (D) occur. For Outcome (C), it follows from (6.7) that x_a and y_a are given by (7.4) provided ρ in these equations is calculated according to (7.5). These expressions are also given in Section 6.4 in terms of a quantity d^* which is just $1/\rho$. Thus, in principle, the determination of the long-term behavior of the general models is no more difficult than it is for Model 1A, once the appropriate version of ρ has been calculated. If seed of all ages softens at the same rate, then, as was shown in Chapter 2, Model 1A is a sub-model of

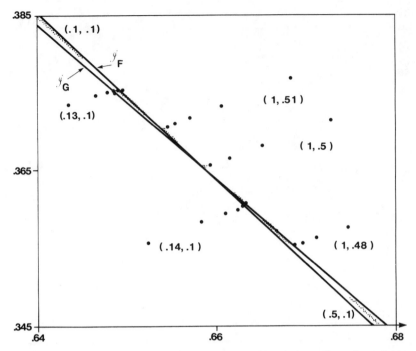

Figure 7.6. As for Figure 7.5 but with window dimensions $0.64 \leqslant p \leqslant 0.68$ and $0.345 \leqslant q \leqslant 0.385$. The orbit starting at $(0.1, 0.1)$ was run for $n \leqslant 200$, $n \leqslant 400$ for $(0.5, 0.1)$, and $\mathbf{n} \leqslant 60$ for all other starting points.

Model 1, and hence the similarity is not so surprising. Indeed, the two versions of ρ given by (7.5) and (3.5) agree in the case of equal softening rates; see also (2.10) and (2.15).

Estimates of rates of convergence to equilibria are also given by Theorems 6.5, 6.10 and 6.11. Under the conditions of Outcome (B), the losing strain vanishes at a rate that is no slower than R_U^n where $R_U = (k_{12}/\rho)^{1/N}$ is interpreted as an upper bounding rate constant. By this we mean that there is a finite constant K such that $x_n \leqslant KR_U^n$ and $|P_V M_2 - y_n| \leqslant KR_U^n$, where P_V is the long-term proportion of volunteer species. The exact rate constant R is given by Theorem 6.7 for the de Wit case $k_{12}k_{21} = 1$, and the discussion near the end of Section 6.4 shows how this is extended to the full generality of Outcome (B). In this case $R = 1/\eta$ where η is the unique solution exceeding unity of (6.12b). In the finite memory case this is a polynomial equation, and for our standard data it is of degree 5 or 6. The constant R_U is easily calculated from the stability criteria and, with modern pocket calculators, the rate constant R is hardly more difficult to compute.

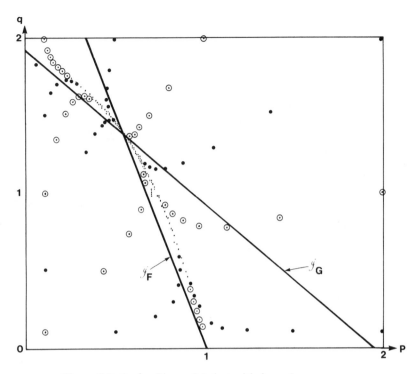

Figure 7.7. As for Figure 7.1, but with $k_{12} = 3$.

In the case of long-term coexistence, Outcome (C) above, we have $k_{12}k_{21} > 1$ and Theorem 6.11 gives $R_U = \theta^{1/N}$ where

$$\theta = 2 - r + (r - 1)^2/rQ, \quad r = k_{12}/\rho \quad \text{and} \quad q = 1 - 1/k_{12}k_{21}.$$

Thus, with x_a and y_a given by (7.4) and (7.5), there is a constant K such that

$$|x_n - x_a| \leqslant KR_U^n \quad \text{and} \quad |y_n - y_a| \leqslant KR_U^n.$$

At the end of Section 6.4 we side-stepped the calculation of the exact rate constant R as being too complicated. Even though Models 1 and 2 are more specific than Model G, a precise derivation of R will be little simpler than the proof given for Model G in Section 5.4. In Section 6.5 we indicated that R can be computed from the linearised analysis there as follows. With a_i and b_i defined as in Section 6.1, define the polynomials $A(t) = \Sigma_{1 \leqslant i \leqslant N} a_i t^i$ and $B(t) = \Sigma_{1 \leqslant i \leqslant N} b_i t^i$, so that $A = A(1)$ and $B = B(1)$. Then $R = 1/\eta$ where η is the unique root exceeding unity of the equation

$$(k_{21}B - A)A(t)/A^2 + (k_{12}A - B)B(t)/B^2 = k_{12}k_{21} - 1.$$

This is equivalent to the equation $D(\eta) = 0$ in Section 6.5.

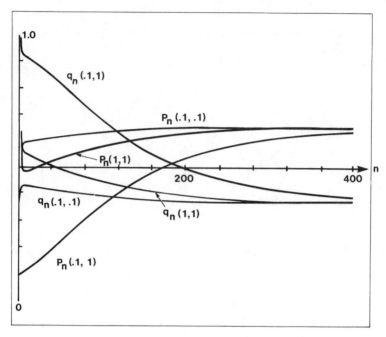

Figure 7.8. p_n and q_n against n for Seaton Park–Midland B with the indicated starting positions, parameter values as in Figure 7.5 except that $n = 1$–400.

Table 7.5. *Stability analysis for Models 1 and 2*

	k_{12}/ρ	$k_{21}\rho$	(x_a, y_a)	R_U	R
Model 1					
Dwal.–Daliak	.5121	1.9527	(0, 87.3)	.895	.761
Yar.–S. Park	.9145	1.0935	(0, 43.2)	.982	.949
S. Park–Mid. B	.9945	1.0055	(0, 59.1)	.999	.997
Model 2					
Dwal.–Daliak	.5090	1.9645	(0, 87.3)	.894	.767
Yar.–S. Park	.9115	1.0971	(0, 43.2)	.982	.949
S. Park–Mid. B	.9920	1.0081	(0, 59.1)	.999	.996
S. Park–Mid.B $k_{12} = 1.07$					
Model 1.	1.0641	1.0055	(39.8, 4.9)	.999	.998
Model 2.	1.0614	1.0081	(38.2, 7.2)	.999	.997

In Table 7.5 we have summarized the long term predictions of Models 1 and 2 for the main mixtures. For the time being we confine our attention to the blocks labelled "Model 1" and "Model 2", where we have set $k_{12} = k_{21} = 1$. The first and second columns show that Outcome (B) occurs for each mixture and each model. We see that the predicted new seed production y_a is very close to that for Model 1A tabulated in the last column of Table 7.4. There are small, but significant, differences in the rate constants R of the various models. Thus Model 1A predicts a slightly slower rate of convergence for the Yarloop–Seaton Park mixture than do Models 1 and 2, but the reverse is true for the other mixtures. Apart from the fact that Models 1 and 2 predict very similar rate constants for each mixture, there is no particular relation between these predictions. Table 7.5 also shows that R_U can be rather larger than R, as it is for the Dwalganup–Daliak mixture. Thus, as was mentioned above, Model 1A gives as good a description of the long-term behavior of these mixtures as do the more general Models 1 and 2. In the next section we will see that this extends to their short-term behavior.

The last two rows of Table 7.5 summarize the long term outcome for the Seaton Park–Midland B mixture when k_{12} is increased from unity to 1.07. As for Model 1A, Outcome (B) gives way to Outcome (C). Models 1 and 2 predict similar long-term new seed productions, although in relative terms Model 2 predicts a 40 percent larger value for Midland B. However both values of y_a are quite small. In both cases the long term new seed productions are quite different to those when $k_{12} = 1$. This occurs because for this mixture the values of ρ listed in Table 7.5 are very close to unity and, in terms of Figure 6.2, an increase of 0.07 in the value of k_{12} represents a comparatively large step off the de Wit hyperbola into Region 3 with correspondingly large changes in x_a and y_a.

A comparison of the last rows of Tables 7.4 and 7.5 shows poor agreement between the long-term predictions of Models 1 and 2 on the one hand, and Model 1A on the other. The reason for this is that the parameters estimated for Model 1A have resulted in a much larger value of $\rho - 1$ than is the case for Models 1 and 2. It is clear from (7.4) that if k_{12} and k_{21} are both close to unity then x_a and y_a are very sensitive to small variations in ρ about unity.

This situation highlights the need for care when assessing the adequacy of the various models. When $k_{12} = 1$ the models all predict Outcome (B) and variations of ρ which keep them within the ambit of this outcome have no effect on predicted new seed production; the models appear to be equivalent in this respect. However, this need not entail equivalences between other aspects of the models. For example, the tabulated values of S_i and T_i for the main mixtures show that for the most part Models 1 and 2 are not well summarized by Model 1A. In order that Model 1A be an acceptable surrogate it is necessary that the S_i and the T_i be roughly in a geometric progression, that is, S_i/S_{i+1} and T_i/T_{i+1} be roughly independent of i. This is true only for the Dwalganup–Daliak mixture, and then only for $1 \leqslant i \leqslant 3$. As we have seen, such differences may become important when variations of other parameters, such as the de Wit constants, induce a transition from Outcomes (A) or (B) to Outcome (C). For the main mixtures, we can safely conclude that Model 1A is adequate when Outcome (B) occurs but if Outcome (C) behavior is indicated then the more general models probably should be used.

7.4   Dominance change in Model 1A with parameter variation

The effect on dominance of changing summer survival and softening rates in the de Wit case of Model 1A can be analysed as follows. In this case we have $k_{12}k_{21} = 1$ and Strain 2 wins if $k_{12}/\rho < 1$, and Strain 1 wins if this inequality is reversed. We begin by examining the ratio

$$k_{12}/\rho = \frac{aP_S P(S)[1 - Q_S + Q_S Q(S)]}{Q_S Q(S)[1 - P_S + P_S P(S)]}$$

where $a = k_{12} W M_1 P_E P_A / M_2 Q_E Q_A$; see Section 3.1 for the notation. This ratio is a monotonic function of $P(S)$ and of $Q(S)$, and Strain 2 wins iff

$$Q_S Q(S)[1 - P_S + P_S P(S)] > aP_S P(S)[1 - Q_S + Q_S Q(S)]$$

or equivalently, iff

$$Q_S Q(S)[1 - P_S - (a - 1)P_S P(S)] > aP_S P(S)(1 - Q_S).$$

This condition cannot hold if $1 - P_S - (a - 1)P_S P(S) \leqslant 0$ and hence Strain 1 will win in this case. Suppose then that this expression is positive, that is,

$$1 - P_S > (a - 1)P_S P(S). \tag{7.6}$$

Under this condition, Strain 2 wins iff

$$Q(S) > \frac{aP_S P(S)(1 - Q_S)}{Q_S[1 - P_S - (a - 1)P_S P(S)]}.$$

Let $f(P(S))$ denote the right-hand side of this inequality. The function $f(.)$ is defined on $[0, 1]$, with the possible exception of a single point, $f(0) = 0$, and it is increasing within the intervals in which it is defined. It may have two branches on $[0, 1]$, but under assumption (7.6) only the positive branch is relevant to our discussion. We consider two cases.

Assume first that $a < 1$. Then (7.6) is true and the function $f(.)$ is defined and concave on $[0, 1]$. Its maximum value is $m_p = f(1)$ which is given by

$$m_P = aP_S(1 - Q_S)/Q_S(1 - aP_S),$$

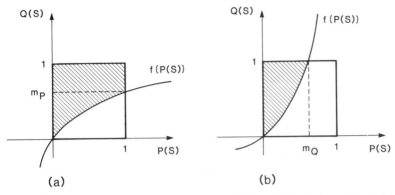

Figure 7.9. Schematic illustration of $f(P(S))$ when (a) $a \leqslant 1$ and (b) $a > 1$. Strain 2 wins in the shaded areas and Strain 1 wins in the unshaded areas.

and this can take any positive value. The case where $m_P < 1$ is illustrated schematically in Figure 7.9a where the shaded area is that portion of the unit square where $Q(S) > f(P(S))$, that is, it shows those combinations of $P(S)$ and $Q(S)$ allowing Strain 2 to win.

Now suppose that $a > 1$. In this case $f(.)$ is defined and convex on $[0, 1]$ provided $aP_S < 1$, but if $aP_S \geqslant 1$ then $f(.)$ is defined and convex only on $[0, (1 - P_S)/(a - 1)P_S]$ and $f(.)$ increases to infinity as its argument approaches the right-hand end point of this interval. If the first case holds and $m_P \leqslant 1$ we have a situation similar to that illustrated in Figure 7.9a, except that the curve is convex. If either $m_P > 1$ or $aP_S \geqslant 1$ then the equation $f(m_Q) = 1$ has a solution in $(0, 1]$ given by

$$m_Q = Q_S(1 - P_S)/P_S(a - Q_S)$$

and which can assume any positive value. This case is illustrated in Figure 7.9b and, as above, the shaded region delineates the domain of dominance by Strain 2.

We can now ask how dominance changes with variations in a, P_S and Q_S. We will restrict our discussion to variations in P_S and Q_S. When $a < 1$ we can write

$$m_P = a(Q_S^{-1} - 1)/(P_S^{-1} - a)$$

and it is clear that m_P is a decreasing function of Q_S and an increasing function of P_S. So, as Q_S decreases toward 0, m_P increases, and the shaded region of Strain 2 dominance in Figure 7.9a shrinks. The same occurs if P_S increases toward 1. Similar conclusions can be drawn from Figure 7.9b by considering m_Q when $a > 1$.

The values of a and m_P for the standard data are as follows:

	a	m_P	$P(S)$	$Q(S)$
Dwalganup–Daliak	.586	.138	.3975	.3975
Yarloop–Seaton Park	1.058	.884	.5025	.5025
Seaton Park–Midland B	1.0014	.686	.5025	.3725
Dwalganup–Northam A	.594	.208	.3975	.3375
Daliak–Dinninup	.563	.143	.4225	.4125
Geraldton–Dinninup	.403	.081	.3525	.4125

The function $f(P(S))$ is plotted against $P(S)$ for the three main mixtures in Figure 7.10a. This function is concave for Dwalganup–Daliak ($a < 1$), convex for Yarloop–Seaton Park ($a > 1$) and almost a straight line for Seaton Park–Midland B ($a \approx 1$). As a result of the concavity of f, dominance is almost independent of the softening rate of Dwalganup in the first mixture, provided the softening rate of Daliak remains above about 0.14. This is not the case for the other two mixtures, where dominance can be changed by changes in softening rate. The standard values of $P(S)$ and $Q(S)$ (listed above) are also plotted in Figure 7.10a. Note that Yarloop–Seaton Park and Seaton Park–Midland B points are very close to the equilibrium line, and only small changes would be required to alter dominance.

In Figure 7.10b we look at the effect on $f(P(S))$ of varying the summer survival rates P_S and Q_S of Yarloop and Seaton Park. Changes of the order of 20 percent in these drastically alter the dominance. Note that the analysis of the last two paragraphs refers only to dominance and reveals nothing about how long the losing strain may persist.

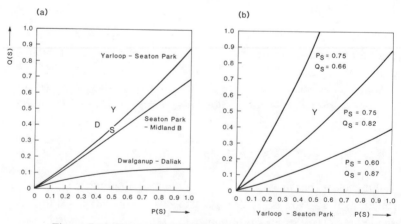

Figure 7.10. Plots of $f(P(S))$ for (a) the three main mixtures and (b) Yarloop–Seaton Park, when P_S and Q_S vary as indicated. The middle curve respresents the standard data values.

However $\Delta = \alpha + (1 - \alpha)k_{12}/\rho$ measures the rate of convergence, and we have $\Delta < 1$ if and only if $k_{12}/\rho < 1$; furthermore Δ is close to 1 if and only if k_{12}/ρ is close to 1. Thus mixtures with k_{12}/ρ close to 1 converge slowly to their equilibria, and rapid convergence occurs when k_{12}/ρ is far from 1. Since the proximity of $Q(S)$ to $f(P(S))$ measures the proximity of k_{12}/ρ to 1, the analysis of the last two paragraphs, and Figures 7.10a and 7.10b, give some idea of the rates of convergence of mixtures with varying summer survival and softening rates.

7.5   Comparison of models and parameter variations

Figures 7.11a–7.11c show the trends in established plants calculated from all the models. Each mixture was assumed to start with an application of new seed in year 1 calculated from the de Wit formula for a 50–50 ratio of adult plants on a weight basis. The characteristic peaked shape of the curve for the dominant strain is due to the time-varying nature of the volunteer species intrusion, which begins in year five as 10 percent of the pasture and rises to 70 percent by year eight, after which it remains at 70 percent (see Table 7.1). In Figure 7.11 the curves for Models 1, 2A and 2 are not separated since they are virtually identical, but they differ somewhat from those for Model 1A, especially in the case of Seaton Park–Midland B. Model 1A describes best the experimental data for this mixture (see Rossiter et al. (1985)).

According to Table 7.5, $k_{12}/\rho < 1$ and $k_{21}\rho > 1$ are satisfied for each mixture for all models for the standard data and so the second strain in each mixture is predicted to be the dominant one. Table 7.5 also lists values of $R = 1/\eta$ and its upper bound R_U. As is borne out by Figure 7.11, all models predict that Dwalganup is eliminated rapidly from the Dwalganup–Daliak mixture, but that Seaton Park and Midland B coexist for well over 100 years; Yarloop and Seaton Park are intermediate between the two.

To display the effects of parameter variation we are restricted to two dimensions so usually we let one pair of parameters, (P_S, Q_S) for example, vary over a range, while holding the remaining parameters from Model 1A fixed at their standard values. But Figure 7.12 shows the trends obtained when P_S (summer survival) and $P(S|B)$ (burr softening rate) are varied over ± 20 percent of their standard values, keeping the differences $P_S - Q_S$ and $P(S|B) - Q(S|B)$ constant. The range ± 20 percent is about two standard errors, according to the data of Rossiter et al. (1985). With this kind of variation the numbers of plants in the pasture change in expected directions (more plants with higher softening rates and/or better summer survival) but the dominance of the strains is not altered.

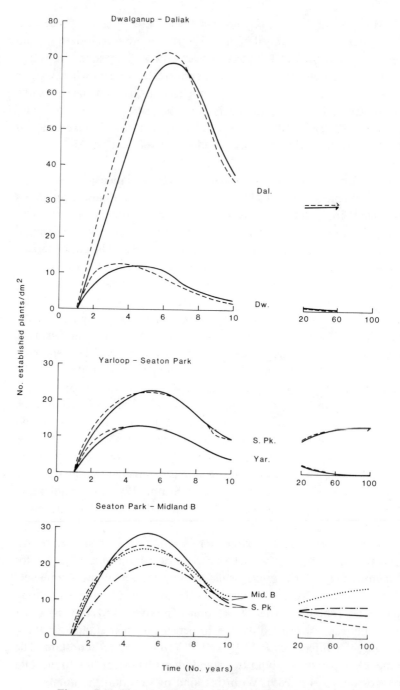

Figure 7.11. Time varying trends for Model 1A (——) and Models 1, 2A and 2 (– – –) (not separated by the accuracy of the diagram). (a) Dwalganup–Daliak, (b) Yarloop–Seaton Park, (c) Seaton Park–Midland B.

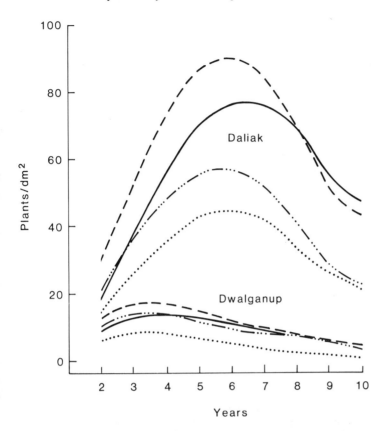

Figure 7.12. Varying $P(S)$ and $Q(S)$, the softening probabilities, and P_S and Q_S, the summer survival rates, for Dwalganup–Daliak in Model 1A. $---$ $(P_S, Q_S) = (.90, .95)$, $(P(S), Q(S)) = (.55, .55)$. $\underline{\qquad}$ $(P_S, Q_S) = (.90, .95)$, $(P(S), Q(S)) = (.39, .39)$. $-\cdots-$ $(P_S, Q_S) = (.66, .71)$, $(P(S), Q(S)) = (.55, .55)$. $\cdots\cdots$ $(P_S, Q_S) = (.66, .71)$, $(P(S), Q(S)) = (.39, .39)$.

The dominance of the strains can alter, however, when the components of a parameter pair such as (P_S, Q_S) are varied separately. To display computer output from such an analysis we use a contour diagram showing the number of years for the elimination of the first strain of a pair (in ordinary print) or the second strain (in italics). Actually, since no population size is ever zero in our models, we adopt the arbitrary convention that a strain is deemed to be eliminated if its population size drops below 0.01 seeds. The contour diagram for summer survival of Seaton Park–Midland B for Model 1A (Figure 7.13a) shows that the standard data are located very close to the equilibrium line in a region of rapid change; that is, small changes in P_S and/or Q_S may lead to changes in dominance. Similar situations may obtain for establishment

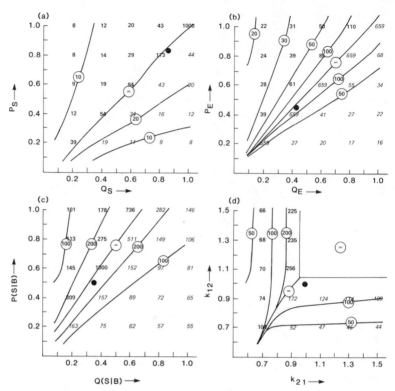

Figure 7.13. Contour plots of the number of years for Seaton Park (roman type) or Midland B (italics) to fall below 0.01 seeds/dm². In each figure the heavy dot shows the standard data value. (a) Varying summer survival (P_S, Q_S). (b) Varying establishment rates (P_E, Q_E). (c) Varying softening rates ($P(S)$, $Q(S)$). (d) Varying de Wit constants (k_{12}, k_{21}). In (d) the de Wit case $k_{12}k_{21} = 1$ is shown by a dotted line. The rectangle marked '∞' is the region of stable coexistence.

rates (Figure 7.13b) and softening rates (Figure 7.13c), though changes occur progressively less rapidly.

In Figure 7.13d we see the effect of varying the de Wit constants k_{12} and k_{21} over the ranges [.6, 1.5], that is, from 60 percent to 150 percent of their standard values. The starting point was a 50/50 mixture. The standard data ($k_{12} = 1 = k_{21}$) is located very close to the junction of the equilibrium and to the region of true coexistence. With the standard data Midland B takes 580 years to reduce Seaton Park to 0.01 seeds. A decrease in k_{21} for Midland B from 1 to 0.9, holding k_{12} at 1, would reverse dominance, Seaton Park then winning in 250 years. The mixture here is in the ambit of Case (D), that is, there is an unstable interior equilibrium. Thus it is important to specify the initial configuration. On the other hand, holding k_{21} at 1 and increasing k_{12} to 1.07 leads to

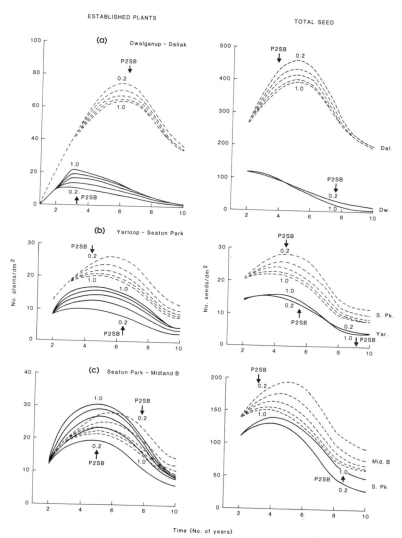

Figure 7.14. (a) Time trends for varying the rate of softening of 2-year-old seed, $P_2(S|B)$, denoted P2SB, in Model 1B for Dwalganup–Daliak. The direction of change of $P_2(S|B)$ is signified by arrows. [$P_2(S|B)$ = 0.2, 0.4, 0.6, 0.8, 1.0]. (b) As in (a), but for Yarloop–Seaton Park. O—O = Yarloop, - - - = Seaton Park. [Note the reversal in total seed numbers for $P_2(S|B)$ = .2 and 1.0.] (c) As in (a), but for Seaton Park–Midland B. O—O = Seaton Park, - - - = Midland B.

stable coexistence for the two strains. Changes of these magnitudes are well within the standard errors of k_{12} and k_{21}.

To describe the effect of differential softening rates we introduce Model 1B, which is the variation of Model 1A in which seed greater

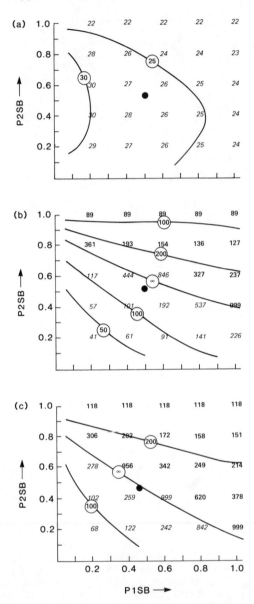

Figure 7.15. Contour plots of the number of years for the first strain (roman type) or the second strain (italics) of each of the main mixtures to fall below 0.01 seeds/dm² in Model 1B when the rates of softening of 1 and 2 year old seed, $P_1(S|B)$ and $P_2(S|B)$, denoted P1SB and P2SB, are varied from 0.1 to 1.0. The heavy dots show where Model 1B reduces to Model 1A (standard data). (a) Dwal.–Dal. (b) Yar.– S.P. (c) S.P.–Mid. B.

than one-year-old softens at a rate which may differ from that of one-year-old seed. We let $P_1(S \mid B)$ and $Q_1(S \mid B)$ be the softening rates of one-year-old burr seed for the two strains; $P_2(S \mid B)$ and $Q_2(S \mid B)$ are the rates for seed aged more than one year. Similarly, $P_1(S \mid F)$, $Q_1(S \mid F)$, $P_2(S \mid F)$ and $Q_2(S \mid F)$ denote the corresponding rates for free seed. We recover Model 1A by setting $P_1(S \mid B) = P_2(S \mid B)$, and so on. Model 1B is a special case of Model 1 (and hence its behavior is described by the analysis of Chapter 6) which is more tractable, and a version of it will be used for the single strain model to be introduced in Chapter 8.

Figures 7.14(a)–7.14(c) show the effect of varying $P_2(S \mid B)$ on the time trends for established plants in the three main mixtures, all other parameters (including $Q_2(S \mid B) = Q_1(S \mid B)$, $Q_2(S \mid F) = Q_1(S \mid F)$) being held at their standard values. The effect is to alter the height of the peaks and possibly also the long-term equilibrium values. (The position of the peaks, as remarked earlier, is dictated by the volunteer species intrusion and cannot be changed by altering the other parameters.) To further investigate the reversal of dominance, in Figure 7.15 we give the contour diagrams (years to extinction) for varying $P_1(S \mid B)$ against varying $P_2(S \mid B)$ for the three main mixtures.

Altering $P_2(S \mid B)$ has a curious effect on the time trends for the maximum seed pool for each strain, as is shown in Figure 7.14 for the three main mixtures. As $P_2(S \mid B)$ changes from 0.1 to 1.0 the total seed for the losing strain stays fairly constant but that for the winning strain changes by almost 20 percent. Changing $Q_2(S \mid B)$ has the reverse effect.

8

A single strain model with cropping and tillage

8.1 Introduction

Previous chapters of this book have dealt in some detail with a mixture of two strains of subterranean clover grown under a "continuous pasture" system, whereby the clover strains, having been established in year one, continue to regenerate in each subsequent year. Our models for mixtures were designed to be applicable to a relatively high rainfall area where such pastures are common, and in particular assumed density independent growth of the subclover under normal grazing conditions, as described in Rossiter et al. (1985). One or both clovers persist indefinitely in these models; there is no question of the clover component of the pasture as a whole dying out.

But in Australia, especially, subterranean clover based pastures are also grown in rotation with crops, and over a wide variety of environmental conditions. In such systems, the clover provides a high quality forage, while maintaining or improving soil fertility via nitrogen fixation for the benefit of the following crop. The lengths of the pasture and crop phases vary with economic and managerial considerations. In this chapter we formulate and analyse models for such systems, restricting ourselves to a "single strain" model, rather than attempting to model binary mixtures. Such models have been applied to agronomic problems in Western Australia by Taylor, Rossiter and Maller (1989).

Consider a model in which a single strain of subclover is sown in year one and regenerates in subsequent years via its residual hard seed pool. If no cropping is imposed we call this a "continuous pasture" system, and a flow diagram for the simplest description of its life cycle would be given by Figure 1.1 of Chapter 1. This is a "single strain" version of Model 1A, where all residual hard seed is pooled and no attempt is made to record its age or distinguish between free and burr seed. (More generally, the model which will be formulated and analysed in this chapter is actually a version of the slightly more complex Model 1B introduced in Chapter 7).

If we impose a "cropping system" whereby the pasture is removed in designated years and a crop (such as wheat or lupins) is sown for that year (or for a number of years in succession) we can describe the subsequent regeneration of the clover, via the residual hard seed pool, on cessation of cropping. Pasture may be removed in two ways: firstly by a minimum tillage system in which weeds are killed by a herbicide and soil disturbance is kept to a minimum (we call this "no tillage" cropping); secondly by a system in which the soil is thoroughly ploughed, perhaps up to four times, to kill weeds and prepare seed beds. This will be our "tillage" system. The procedure used in practice depends on environmental conditions such as soil type, as well as economic considerations.

A number of new aspects arise if we wish to model these processes in a realistic way. First is the fact that burial of seed in the soil affects the rate of seed softening. Seed buried and redistributed by the tillage process includes any hard seed present at the surface together with hard seed remaining below the surface as a consequence of previous tillage operations. A second aspect is that to adequately describe subclover regeneration over the range of environments to which we wish to apply our models, we must assume seed production to be density dependent. This is necessary in the single strain model (whereas it was not in the mixture model) due to the lower plant densities which may occur in low rainfall regions, for example. We model this aspect by introducing a density dependence function at the seed production stage of the cycle. Taylor et al. (1989) give a fuller explanation and examples of the density dependence function.

Taylor (1981, 1984), in a series of field and laboratory experiments, has elucidated a number of surprising aspects of the softening of buried subterranean clover seed. Of especial interest here is his discovery that the rate of seed softening is slowed when seed is held at depths of up to 10 cm in the soil. Seed can be distributed down to this depth by the tillage process. On the other hand, a small proportion of seeds may remain hard for up to seven years when left undisturbed at the soil surface. Survival of seeds over the summer period is also improved by burial, since the seeds are less susceptible to grazing stock and premature germination from summer rains.

A second important discovery of Taylor's is that a proportion of hard seed, once having been buried in the soil, attains a "latent" ability to soften more rapidly after it has been returned to the surface by another tillage operation than does seed which has not had this burial experience. This rapidly softening seed is of great benefit to clover re-establishment in the pasture year following a crop phase.

In this chapter we describe a deterministic model which follows the progress of a single strain of subterranean clover through a variety of pasture/crop regimes, with special allowance made for the processes of seed distribution through the soil by tillage, and of the softening of buried latent soft seed on its return to the vicinity of the surface. Actually, we describe a series of models ranging from continuous pasture through to various relatively complex cyclical cropping schemes with or without tillage. With these models, we are able to investigate the relative importance of various aspects of the life history of the subterranean clover, such as the rate of seed softening, summer survival, and so forth, and aspects of the management regime, for determining whether the pasture "succeeds" or "fails" (dies out), how long it takes to do these things, and the number of plants (hence, the amount of pasture) produced in the meantime.

8.2   General description and notation

The model which we formulate and analyse in this chapter has two main aspects: it follows the growth of the subclover on the surface of the ground; and it considers the processes of cropping and "tillage", by which the seed is distributed underground.

The first aspect can be conveniently separated from the second in the *continuous pasture model*, which is the first to be formulated in the next section. This model is one in which cropping and tillage are absent, so the below-ground component of the model need not be considered. The operation of the model can be seen from the flow chart of Figure 8.1 and is similar to Model 1B defined in Chapter 7. New seed produced in year n plus residual hard seed produced in previous years survives the summer of year n at rate $P_S(0)$. This seed softens at rate $P_1(S)$, while seed more than one-year-old softens at the possibly different rate $P_2(S)$. The soft seed gives rise to established plants at the rate P_E and to adults at rate P_A. The new seed produced in year $n + 1$ is the value of the density dependence function $f(x)$ whose argument is the number of adult plants. All seed not softening is pooled into "residual hard seed at surface".

The second, or below-ground, aspect of the model only comes into consideration when the pasture is tilled in crop years. On the surface, events are assumed to take place as for the continuous pasture, until tillage occurs. In the course of tillage, seed is distributed below the surface in the crop year. We model this process by assuming that proportions $p_1, \ldots, p_d$ of the seed are distributed to depths $D_1, \ldots, D_d$.

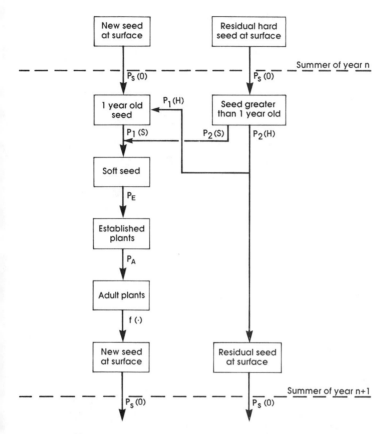

Figure 8.1. Flow diagram for the single strain model as used in this chapter to describe the continuous pasture system.

We assume there is *latent softening* of the seed, by which a proportion of seed at depths D_2, D_3, ... softens during the summer following its redistribution to D_1. Seed germinating below a certain depth is *non-emergent*; only seed from depth D_1 is assumed to emerge as seedlings. The seed production of established plants is assumed, as before, to be density dependent. Not included in the model is the possible effect of age-dependent softening of seed, and all seed is assumed to be burr seed. These restrictions could be relaxed, in a more general model, as for the cases of the mixture models, but this will not be attempted here.

Notation will be as follows. Consider d depths $d \geqslant 2$, denoted D_1, D_2, ..., D_d, and let D_0 denote the surface, that is, zero depth. On tillage, seed is assumed to be distributed among D_1, ..., D_d in the proportions p_1, ..., p_d, $p_1 + p_2 + ... + p_d = 1$. Let $P_S(D_i)$ $i = 0, 1, ..., d$, be the rate at which seed at depth D_i survives summer, $P(S \mid D_i)$, $i = 0, 1, ..., d$ the rate at which seed at depth D_i softens, $P_i(L \mid H)$, $i = 2, 3, ..., d$, the rate at which seed at depth D_i becomes latent soft. (We assume no seed becomes latent soft at depth 1, so $P_1(L|H)$ is not required.) Also let $P(H|D_i) = 1 - P(S|D_i)$, $i = 0, 1, \ldots, d$, and $P_i(H|H) = 1 - P_i(L|H)$, $i = 2, 3, \ldots, d$. As before let P_E be the probability that a seed establishes an adult plant, and let $f(.)$ denote the seed production function.

Reference to the flow diagram of Figure 8.2 shows the operation of the model. For simplicity only three depth intervals are shown, $D_1 = 0 - 2.5\,\text{cm}$, $D_2 = 2.5 - 5.0\,\text{cm}$, $D_3 = 5.0 - 7.5\,\text{cm}$; these being the values used in Taylor et al. (1989). Of course $D_0 = 0\,\text{cm}$. The diagram shows what happens in a crop year with tillage, preceded and succeeded by pasture years. The model operates as follows: seed at depths 0, 2.5, 5.0, 7.5 cm survives the summer following a pasture year with probabilities $P_S(0)$, $\ldots$, $P_S(7.5)$. At the surface, some of the seed is new, having been produced in the pasture year, while some may be residual hard seed more than one-year-old. A transition from hard to latent soft seed takes place at 2.5, 5.0 and 7.5 cm. At the surface new seed softens with probability $P_1(S|0)$ while seed greater than one-year-old softens with probability $P_2(S|0)$. At depth 2.5 cm, seed softens with probability $P(S|2.5)$; there will be no latent soft seed at this stage since it all softened during the previous (pasture) year. Seed softening from depths 5.0 and 7.5 with probabilities $P(S|5.0)$, $P(S|7.5)$, is non-emergent and is lost.

On tilling, the hard seed from all depths is mixed together and redistributed among depths 2.5, 5.0 and 7.5 cm in proportions p_1, p_2 and p_3. (Similarly latent soft seed, if present, would be mixed and redistributed among these depths in the same proportions). Plants produced from emergent soft seed are lost to the tilling process, and produce no seed.

Following another summer's losses, seed comes into the pasture year, undergoes a transition from hard to latent soft, and softens as in the previous year. Emergent seed now gives rise to established plants with probability P_E and adult plants with probability P_A, which produce seed according to the function $f(.)$. (A box is included on the flow diagram to denote possible residual hard seed at the surface which would only be present if a pasture or no tillage crop year had preceded.) All hard seed remaining now simply passes into the next year, after another summer's losses.

8.3   Mathematical formulation

(A)   *Continuous pasture system*

To formulate the model described in the introduction, let x_n be the new seed produced in year n, r_n the residual hard seed in year n and

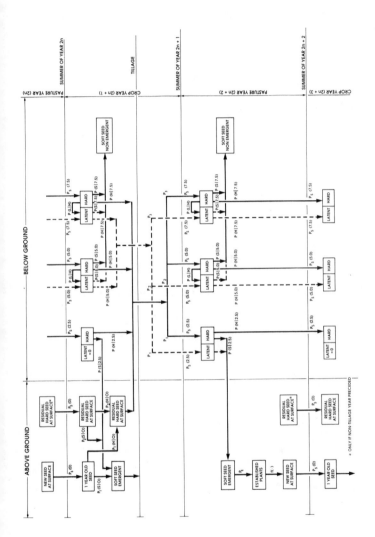

Figure 8.2. Flow diagram for the cropping/tillage model.

e_n the established plants in year n. In year $n + 1$ the new seed produced will then be

$$x_{n+1} = f\{P_S(0)P_1(S)P_E P_A x_n + P_S(0)P_2(S)P_E P_A r_n\}$$
$$= f(t_1 x_n + t_2 r_n), \text{say},$$

remembering that new seed softens at rate $P_1(S)$ while older seed softens at rate $P_2(S)$. Here

$$t_1 = P_S(0)P_1(S)P_E P_A, \quad t_2 = P_S(0)P_2(S)P_E P_A.$$

Now, taking $r_1 = 0$,

$$r_n = P_S(0)[P_1(H)x_{n-1} + P_2(H)r_{n-1}]$$
$$= P_S(0)P_1(H)x_{n-1} + P_S(0)P_2(H)\{P_S(0)[P_1(H)x_{n-2} + P_2(H)r_{n-2}]\}$$
$$\vdots$$
$$= P_1(H) \sum_{j=0}^{n-2} P_S^{j+1}(0)P_2^j(H)x_{n-j-1}. \tag{8.1}$$

Substituting gives a difference equation for x_{n+1}:

$$x_{n+1} = f\{t_1 x_n + t_2 P_1(H) \sum_{j=0}^{n-2} P_S^{j+1}(0)P_2^j(H)x_{n-j-1}\}. \tag{8.2}$$

Since $x_n = f(P_A e_n)$ we also have a difference equation for e_{n+1}:

$$e_{n+1} = P_S(0)P_1(S)P_E x_n + P_S(0)P_2(S)P_E r_n$$
$$= P_S(0)P_1(S)P_E f(P_A e_n)$$
$$+ P_2(S)P_E P_1(H) \sum_{j=0}^{n-2} P_S^{j+1}(0)P_2^j(H)f(P_A e_{n-j-1}). \tag{8.3}$$

These equations simplify in two special cases: firstly if the softening rate of new and old seed is the same, $P_1(S) = P_2(S) = P(S)$, say, and $P(H) = 1 - P(S)$, then

$$e_{n+1} = P_S(0)P(S)P_E f(P_A e_n) + P_S(0)P(H)e_n \tag{8.4}$$

as is easily checked. Secondly, when yield does not depend on input, that is, $f(.) = f = $ constant, we can solve the whole system explicitly:

$$x_{n+1} = f;$$

$$r_{n+1} = f P_1(H)P_S(0)\frac{[1 - P_S^n(0)P_2^n(H)]}{[1 - P_S(0)P_2(H)]}; \tag{8.5}$$

$$e_{n+1} = f P_S(0)P_1(S)P_E + f P_S^2(0)P_2(S)P_E \frac{[1 - P_S^{n-1}(0)P_2^{n-1}(H)]}{[1 - P_S(0)P_2(H)]}. \tag{8.6}$$

(B)   *Cropping models without tillage*

We assume a regular cropping regime in which seed production in an equally spaced sequence of years is zero, but in the remaining years takes place as in the continuous pasture model. The alternate years cropping system, in which years 1, 3, 5, ... are pasture years whereas seed production in years 2, 4, 6, ... is zero, is simple to formulate and displays the main ideas, so we set it out explicitly before giving a more general formulation.

We have seed production of zero in even years, that is, $x_{2n} = 0$. In year $2n + 1$,

$$x_{2n+1} = f(t_2 r_{2n}),$$

$$
\begin{aligned}
r_{2n+1} &= P_S(0)[0 + P_2(H) r_{2n}] \\
&= P_S(0) P_2(H) P_S(0)[P_1(H) x_{2n-1} + P_2(H) r_{2n-1}] \\
&= P_S^2(0) P_2(H) P_1(H) x_{2n-1} + P_S^2(0) P_2^2(H) P_S(0)[0 + P_2(H) r_{2n-2}] \\
&= P_S^2(0) P_2(H) P_1(H) x_{2n-1} + P_S^3(0) P_2^3(H) r_{2n-2} \\
&\quad \vdots \\
&= P_1(H) \sum_{j=0}^{n-1} P_S^{2j+2}(0) P_2^{2j+1}(H) x_{2n-2j-1} + P_S^{2n-1}(0) P_2^{2n-1}(H) r_2.
\end{aligned}
$$

(8.7)

For established plants, we are only interested in the values in pasture years, in which case

$$
\begin{aligned}
e_{2n+1} &= 0 + P_S(0) P_2(S) P_E r_{2n} \\
&= P_S(0) P_2(S) P_E P_S(0)[P_1(H) x_{2n-1} + P_2(H) r_{2n-1}] \\
&= P_S^2(0) P_1(H) P_2(S) P_E f(P_A e_{2n-1}) \\
&\quad + P_2(S) P_E P_1(H) \sum_{j=0}^{n-2} P_S^{2j+4}(0) P_2^{2j+2}(H) f(P_A e_{2n-2j-3}) \\
&\quad + P_2(S) P_E P_S^{2n+1}(0) P_2^{2n}(H) r_2 \\
&= P_E P_2(S) P_1(H) \sum_{j=0}^{n-1} P_S^{2j+2}(0) P_2^{2j}(H) f(P_A e_{2n-2j-1}) \\
&\quad + P_2(S) P_E P_S^{2n+1}(0) P_2^{2n}(H) r_2.
\end{aligned}
$$

(8.8)

These formulae display, as expected, a cycle of length 2.

We now formulate a more general cropping system, still without tillage. Suppose there are N pasture years followed by M crop years, the pattern being repeated indefinitely. Consider the pasture years immediately following a crop year:

$$r_{n(N+M)+1} = P_S(0)[0 + P_2(H)r_{n(N+M)}];$$

$$r_{n(N+M)+2} = P_S(0)[P_1(H)x_{n(N+M)+1} + P_2(H)r_{n(N+M)+1}]$$
$$= P_S(0)P_1(H)x_{n(N+M)+1} + P_S^2(0)P_2^2(H)r_{n(N+M)};$$

$$\vdots$$

$$r_{n(N+M)+N} = P_1(H) \sum_{j=1}^{N-1} P_S^{N-j}(0)P_2^{N-j-1}(H)x_{n(N+M)+j}$$
$$+ P_S^N(0)P_2^N(H)r_{n(N+M)}. \tag{8.9}$$

Continuing, we follow the residual hard seed through the next M crop years:

$$r_{n(N+M)+N+1} = P_S(0)P_1(H)x_{n(N+M)+N} + P_S(0)P_2(H)r_{n(N+M)+N}$$

$$r_{n(N+M)+N+2} = 0 + P_S(0)P_2(H)r_{n(N+M)+N+1}$$
$$= P_S(0)P_1(H)P_2(H)x_{n(N+M)+N} + P_S^2(0)P_2^2(H)r_{n(N+M)+N}$$

$$\vdots$$

$$r_{n(N+M)+N+M} = P_S^M(0)P_1(H)P_2^{M-1}(H)x_{n(N+M)+N}$$
$$+ P_S^M(0)P_2^M(H)r_{n(N+M)+N}$$
$$= P_S^M(0)P_1(H)P_2^{M-1}(H)x_{n(N+M)+N}$$
$$+ P_S^M(0)P_2^M(H)\{P_1(H) \sum_{j=1}^{N-1} P_S^{N-j}(0)P_2^{N-j-1}(H)x_{n(N+M)+j}$$
$$+ P_S^N(0)P_2^N(H)r_{n(N+M)}\},$$
$$= P_S^M(0)P_2^M(H)P_1(H) \sum_{j=1}^{N} P_S^{N-j}(0)P_2^{N-j-1}(H)x_{n(N+M)+j}$$
$$+ P_S^{N+M}(H)P_2^{N+M}(H)r_{n(N+M)} \tag{8.10}$$

where in the second last step we substituted from (8.9). Equation (8.10) is a recursion relation of the following type for $r_{n(N+M)}$:

$$r_{(n+1)(N+M)} = a_n + P_S^{N+M}(0)P_2^{N+M}(H)r_{n(N+M)}$$

which is easily solved to yield

$$r_{(n+1)(N+M)} = P_1(H) \sum_{i=0}^{n-1} \sum_{j=1}^{N} P_S^{(i+1)(N+M)-j}(0)P_2^{(i+1)(N+M)-j-1}(H)x_{(n-i)(N+M+j)}$$
$$+ P_S^{n(N+M)}(0)P_2^{n(N+M)}(H)r_{N+M} \tag{8.11}$$

which is valid for $N \geqslant 1$, $M \geqslant 1$.

The residual hard seed for the remaining years can now be calculated from:

$$r_{(n+1)(N+M)+k} = P_S^k(0)P_2^k(H)r_{(n+1)(N+M)}$$
$$+ P_1(H)\sum_{j=1}^{k-1} P_S^{k-j}(0)P_2^{k-j-1}(H)x_{(n+1)(N+M)+j},$$

(8.12)

for $1 \leqslant k \leqslant N$, where $r_{(n+1)(N+M)}$ is given by (8.11), and

$$r_{(n+1)(N+M)+N+m} = P_S^M(0)P_2^M(H)r_{(n+1)(N+M)+N}$$
$$+ P_S^m(0)P_1(H)P_2^{m-1}(H)x_{(n+1)(N+M)+N}, \quad (8.13)$$

for $1 \leqslant m \leqslant N$, where $r_{(n+1)(N+M)+N}$ is given by (8.12) with $k = N$.

Now for established plants: we have $x_n = 0$ except when n has the form $n(M + N) + k$, $1 \leqslant k \leqslant N$, and then

$$x_{n(N+M)+k} = f(P_A e_{n(N+M)+k}).$$

Thus in the first pasture year following a succession of M crop years,

$$e_{n(N+M)+1} = 0 + P_S(0)P_2(S)P_E r_{n(N+M)}$$
$$= P_S(0)P_2(S)P_E P_1(H)\sum_{i=0}^{n-1}\sum_{j=1}^{N} P_S^{(i+1)(N+M)-j}(0)$$
$$\times P_2^{(i+1)(N+M)-j-1}(H)f(P_A e_{(n-i-1)(N+M)+j})$$
$$+ P_S^{n(N+M)+1}(0)P_2^{n(N+M)}(H)P_2(S)P_E r_{N+M}. \quad (8.14)$$

For the $(N - 1)$ remaining pasture years before the next crop year, we have

$$e_{n(N+M)+k+1} = P_S(0)P_1(S)P_E x_{n(N+M)+k} + P_S(0)P_2(S)P_E r_{n(N+M)+k}$$
$$= P_S(0)P_1(S)P_E f(P_A e_{n(N+M)+k}) + P_S(0)P_2(S)P_E P_1(H)$$
$$\times \sum_{j=1}^{k-1} P_S^{k-j}(0)P_2^{k-j-1}(H)f(P_A e_{n(N+M)+j})$$
$$+ P_S^k(0)P_2^k(H)e_{n(N+M)+1}$$
$$+ P_S^{n(N+M)+k}(0)P_2^{n(M+M+k)}(H)r_{N+M} \quad (8.15)$$

where $1 \leqslant k < N$.

These formulae are not intended for the calculation of the time dependent quantities x_n, r_n, e_n. This is much more easily done in a recursive fashion on a computer. However we need the explicit representations for asymptotic analysis of the various cropping regimes.

To conclude this section we remark that we have not exhausted all possible cropping regimes of interest – for example, a system with cropping in years 2, 4, 6, ... *and* in years 7, 13, 19, ... Rather than attempt a general formulation to cover such setups, however, we have to analyse them separately at present.

(C)   *Alternating pasture-crop system with tillage*

Suppose even years are pasture years, odd years are crop years. Starting from year $2n + 2$, we follow the fate of the seed. Let $H_{2n+2}(D_i)$, $1 \leqslant i \leqslant d$, be the hard seed surviving the summer of year $2n + 1$ before the transition to latency and before softening. Let $L_{2n+2}(D_i)$, $1 \leqslant i \leqslant d$, be the latent seed at the same stage. Then because the tillage operation mixes the seed of all depths together equally,

$$\frac{H_{2n+2}(D_i)}{P_S(D_i)p_i}$$

are all the same, equal to H_{2n+2}, say. Similarly

$$\frac{L_{2n+2}(D_i)}{P_S(D_i)p_i} = L_{2n+2}, \quad 1 \leqslant i \leqslant d.$$

Clearly H_{2n+2} and L_{2n+2} are the total hard and latent seed pools in year $2n + 1$ before summer losses and before being redistributed among the d depths.

Now seed at depths $D_2, D_3, \ldots, D_d$ may become latent, but not seed at depth D_1, so the latent and hard seed after summer losses and after this transition are:

Latent: $P_S(D_1)p_1 L_{2n+2}$ for D_1

             $P_S(D_i)p_i[L_{2n+2} + P_i(L|H)H_{2n+2}]$ for $D_i, 2 \leqslant i \leqslant d$

Hard $P_S(D_1)p_1 H_{2n+2}$ for D_1

             $P_S(D_i)p_i P_i(H|H)H_{2n+2}$ for $D_i, 2 \leqslant i \leqslant d$

where $P_i(H|H) = 1 - P_i(L|H)$ is the proportion of hard seed remaining hard, that is, not becoming latent.

Emergent soft seed only occurs from D_1. Here all latent seed softens and $P(S|D_1)$ of the hard seed softens, so

emergent soft seed in year $2n + 2$

$$= p_1 P_S(D_1)L_{2n+2} + p_1 P_S(D_1)P(S|D_1)H_{2n+2}$$

$$= p_1 P_S(D_1)[L_{2n+2} + P(S|D_1)H_{2n+2}].$$

This seed establishes with probability P_E and survives to adulthood with probability P_A. Thus for new seed produced in year $2n + 2$, x_{2n+2}, we have

$$x_{2n+2} = f\{P_E P_A p_1 P_S(D_1)[L_{2n+2} + P(S|D_1)H_{2n+2}]\}. \qquad (8.16)$$

Hard seed left after softening in year $2n + 2$ is:

$$p_1 P_S(D_1) P(H | D_1) H_{2n+2} \qquad \text{for } D_1$$

and

$$p_i P_S(D_i) P_i(H | H) P(H | D_i) H_{2n+2} \quad \text{for } D_i, \ 2 \leqslant i \leqslant d,$$

and after a further summer's loss these become $H_{2n+3}(D_i)$, which we define (as above) as the hard seed in year $2n + 3$ before the transition to latency and before softening. Thus

$$H_{2n+3}(D_1) = p_1 P_S^2(D_1) P(H | D_1) H_{2n+2} \qquad \text{for } D_1 \qquad (8.17a)$$

and

$$H_{2n+3}(D_i) = p_i P_S^2(D_i) P_i(H | H) P(H | D_i) H_{2n+2}, \quad \text{for } D_i, \ 2 \leqslant i \leqslant d. \qquad (8.17b)$$

Similarly the latent seed left after softening in year $2n + 2$ is

$$0 \qquad \text{for } D_1 \qquad (8.18a)$$

and

$$p_i P_S(D_i) P(H | D_i) [L_{2n+2} + P_i(L | H) H_{2n+2}] \quad \text{for } D_i, \ 2 \leqslant i \leqslant d, \qquad (8.18b)$$

which after another summer's loss gives $L_{2n+3}(D_i)$, the latent seed in year $2n + 3$ before softening. Thus

$$L_{2n+3}(D_1) = 0 \qquad \text{for } D_1 \qquad (8.19a)$$

and

$$L_{2n+3}(D_i) = p_i P_S^2(D_i) P(H | D_i) [L_{2n+2} + P_i(L | H) H_{2n+2}], \qquad \text{for } D_i, \ 2 \leqslant i \leqslant d. \qquad (8.19b)$$

Now go back to the crop year $2n + 1$. The one-year-old hard seed left from softening but before tillage is $P_1(H | D_0) P_S(D_0) x_{2n}$. There is no residual hard seed left at the surface since this was turned in with the tilling in year $2n - 1$. (This is a peculiarity of the alternating crop–pasture system and does not occur if there are two or more pasture years in a row.) The hard seed aged more than one year (at depths > 0) left after softening and after the transition to latency but before tillage is, by (8.17a),

$$P(H | D_i) H_{2n+1}(D_1) = p_1 P_S^2(D_1) P^2(H | D_1) H_{2n} \quad \text{for } D_1$$

and, by (8.17b)

$$P(H | D_i) P_i(H | H) H_{2n+1}(D_i) = p_i P_S^2(D_i) P^2(H | D_i) P_i^2(H | H) H_{2n}$$

$$\text{for } D_i, \ 2 \leqslant i \leqslant d.$$

Similarly, using (8.18) and (8.19), the latent seed aged more than one year left after softening (and after the transition from hard) but before tillage is

$$0 \qquad \text{for } D_1$$

and

$$P(H|D_i)[L_{2n+1}(D_i) + P_i(L|H)H_{2n+1}(D_i)] = p_i P_S^2(D_i) P^2(H|D_i)$$
$$\times [L_{2n} + P_i(L|H)H_{2n} + P_i(L|H)P_i(H|H)H_{2n}]$$

$$\text{for } D_i, \; 2 \leqslant i \leqslant d.$$

Tillage in year $2n+1$ adds up latent and hard seed separately over all depths including D_0, which, before being distributed in the proportions $p_1, \ldots, p_d$, must amount to H_{2n+2} and L_{2n+2}. Hence

$$H_{2n+2} = P_1(H|D_0)P_S(D_0)x_{2n}$$
$$+ \left[p_1 P_S^2(D_1)P^2(H|D_1) + \sum_{i=2}^{d} p_i P_S^2(D_i)P^2(H|D_i)P_i^2(H|H) \right] H_{2n}$$
$$= aH_{2n} + bx_{2n}, \quad \text{and} \tag{8.20}$$

$$L_{2n+2} = \sum_{i=2}^{d} p_i P_S^2(D_i)P^2(H|D_i)L_{2n}$$
$$+ \sum_{i=2}^{d} p_i P_S^2(D_i)P^2(H|D_i)P_i(L|H)[1 + P_i(H|H)]H_{2n}$$
$$= cL_{2n} + dH_{2n}, \quad \text{say, where we define}$$

$$a = p_1 P_S^2(D_1)P^2(H|D_1) + \sum_{i=2}^{d} p_i P_S^2(D_i)P^2(H|D_i)P_i^2(H|H),$$

$$b = P_1(H|D_0)P_S(D_0),$$

$$c = \sum_{i=2}^{d} p_i P_S^2(D_i)P^2(H|D_i),$$

and

$$d = \sum_{i=2}^{d} p_i P_S^2(D_i)P^2(H|D_i)P_i(L|H)[1 + P_i(H|H)].$$

Solving the second equation for L_{2n} in terms of H_{2n} gives

$$L_{2n} = d \sum_{i=1}^{n-1} c^{n-i-1} H_{2i} + c^{n-1}L_2, \quad n \geqslant 2,$$

while solving the first for H_{2n} in terms of x_{2n} gives

$$H_{2n} = b \sum_{i=1}^{n-1} a^{n-i-1} x_{2i} + a^{n-1}H_2, \quad n \geqslant 2.$$

Substituting in (8.16) gives

$$x_{2n} = f\left\{ P_E P_A p_1 P_S(D_1) \left\{ d \sum_{i=1}^{n-1} c^{n-i-1} H_{2i} + c^{n-1} L_2 + P(S \mid D_1) \right.\right.$$

$$\left.\left. \times \left[b \sum_{i=1}^{n-1} a^{n-i-1} x_{2i} + a^{n-1} H_2 \right] \right\}\right\}$$

$$= f\left\{ P_E P_A p_1 P_S(D_1) \left\{ d \sum_{i=2}^{n-1} c^{n-i-1} \left[b \sum_{j=1}^{i-1} a^{i-j-1} x_{2j} + a^{i-1} H_2 \right] \right.\right.$$

$$\left. + c^{n-2} H_2 + c^{n-1} L_2 + P(S \mid D_1) b \sum_{i=1}^{n-1} a^{n-i-1} x_{2i} + P(S \mid D_1) a^{n-1} H_2 \right\}\right\}$$

$$= f\left\{ P_E P_A p_1 P_S(D_1) \left\{ bd(c-a)^{-1} \sum_{i=1}^{n-3} (c^i - a^i) x_{2(n-i-1)} \right.\right.$$

$$+ ad(c-a)^{-1} H_2(c^{n-2} - a^{n-2}) + c^{n-2} H_2 + c^{n-1} L_2 + P(S \mid D_1) b$$

$$\left.\left. \times \sum_{i=0}^{n-2} a^i x_{2(n-i-1)} + P(S \mid D_1) a^{n-1} H_2 \right\}\right\}$$

which is a difference equation of order n for x_{2n}. Similarly from $H_{2n} = a H_{2n-2} + b x_{2n-2}$ and the expression for L_{2n} in terms of H_{2n} we get

$$H_{2n} = a H_{2n-2} + bf\left\{ P_E p_1 P_A P_S(D_1) \right.$$

$$\left. \times \left[d \sum_{i=1}^{n-1} c^{n-i-1} H_{2i} + c^{n-1} L_2 + P(S \mid D_1) H_{2n-2} \right] \right\}.$$

Finally if e_{2n+2} are the *established plants* in year $2n + 2$, we have, instead of (8.16)

$$e_{2n+2} = P_E p_1 P_S(D_1)[L_{2n+2} + P(S \mid D_1) H_{2n+2}]$$

and instead of (8.20) we have

$$H_{2n+2} = a H_{2n} + bf(P_A e_{2n}),$$

and hence

$$H_{2n} = b \sum_{i=1}^{n-1} a^{n-i-1} f(P_A e_{2i}) + a^{n-1} H_2.$$

Thus we obtain, finally,

$$
\begin{aligned}
e_{2n} &= P_E p_1 P_S(D_1)\left[d \sum_{i=1}^{n-1} c^{n-i-1} H_{2i} + c^{n-1} L_2 + P(S|D_1) H_{2n} \right] \\
&= P_E p_1 P_S(D_1)\left\{ d \sum_{i=2}^{n-1} c^{n-i-1}\left[b \sum_{j=1}^{i-1} a^{i-j-1} f(P_A e_{2j}) + a^{i-1} H_2 \right] + c^{n-2} H_2 \right. \\
&\quad \left. + c^{n-1} L_2 + P(S|D_1) b \sum_{i=1}^{n-1} a^{n-i-1} f(P_A e_{2i}) + P(S|D_1) a^{n-1} H_2 \right\} \\
&= P_E p_1 P_S(D_1)\left\{ db(c-a)^{-1} \sum_{i=1}^{n-3} (c^i - a^i) f(e_{2(n-i-1)}) + ad(c-a)^{-1} \right. \\
&\quad \times H_2(c^{n-2} - a^{n-2}) + c^{n-2} H_2 + c^{n-1} L_2 + P(S|D_1) b \\
&\quad \left. \times \sum_{i=0}^{n-2} a^i f(e_{2(n-i-1)}) + P(S|D_1) a^{n-1} H_2 \right\}.
\end{aligned}
\tag{8.21}
$$

8.4 Asymptotic analysis

(A) Introduction

In this section some asymptotic properties of the previously formulated models are examined. We need to know if the systems reach unique equilibrium points, and if so whether these points are GAS. We are also interested in how the equilibrium points depend on the parameters of the model, such as softening rates. We assume throughout that the density dependence function $f(x)$ is a concave function on $x \geqslant 0$, continuous and with a continuous derivative on $x > 0$, with $f(0) = 0$.

The analysis given in this section does not answer all questions of interest about the asymptotic properties. In particular, systems with tillage are difficult to analyse. But it will be proved that the continuous pasture system under the restriction ((8.25) below) that seed greater than one-year-old softens at least as quickly as new seed, and alternating crop–pasture systems with only one intervening pasture year and without tillage, always possess GAS equilibrium points. That is, in the long run, there is a limiting value to which the number of established plants (and new seed production, residual hard seed, and so forth) in the system converges, and this limiting value is approached from all starting points of the system. We concentrate on established plant numbers in this and the next section since they are of most interest in practice.

As mentioned, there are some gaps in this treatment due mainly to our inability to analyse models with tillage, so it is worth mentioning that in numerous computer simulations that have been made for all these models none, including those incorporating the tillage aspect, has given any reason to suspect behavior other than rapid convergence to GAS equilibrium points.

The range of the present section is broadened, on the other hand, by including models in which seed production varies either cyclically or randomly over the years. While each system has its own peculiarities, both behave mainly in conformity with the other models. These systems are considered in subsections (d) and (e).

Before commencing the study of models, we give a result which enables us to analyse the increments of the models without tillage.

Theorem 8.1.

Suppose a_n is any sequence, $n \geqslant 1$, and let

$$b_{n+1,m} = \sum_{j=0}^{n-m} P^j a_{n-j-1}$$

for $n \geqslant 2$, $0 \leqslant m \leqslant n$, where $b_{n+1,n} = 0$ and $0 < P < 1$. Then

$$b_{n+1,m} - b_{n,m} = a_{n-1} - (1 - P)b_{n,m}, n \geqslant 2. \tag{8.22}$$

Proof. From the definition of $b_{n,m}$,

$$b_{n+1,m} - b_{n,m} = \sum_{j=0}^{n-m-1} P^j (a_{n-j-1} - a_{n-j-2}) + P^{n-m} a_{m-1}$$

$$= \sum_{j=0}^{n-m-1} \sum_{k=j}^{\infty} (P^k - P^{k+1})(a_{n-j-1} - a_{n-j-2}) + P^{n-m} a_{m-1}$$

$$= (1 - P) \sum_{k=0}^{n-m-1} P^k \sum_{j=0}^{k} (a_{n-j-1} - a_{n-j-2})$$

$$+ (1 - P) \sum_{k=n-m}^{\infty} P^k \sum_{j=0}^{n-m-1} (a_{n-j-1} - a_{n-j-2}) + P^{n-m} a_{m-1}$$

$$= (1 - P) \sum_{k=0}^{n-m-1} P^k (a_{n-1} - a_{n-k-2}) + (1 - P)$$

$$\times \sum_{k=n-m}^{\infty} P^k (a_{n-1} - a_{m-1}) + P^{n-m} a_{m-1}$$

$$= (1 - P^{n-m})a_{n-1} - (1 - P) \sum_{k=0}^{n-m-1} P^k a_{n-k-2} + P^{n-m} a_{n-1}$$

$$= a_{n-1} - (1 - P)b_{n,m}.$$

(B) *The continuous pasture model*
This model, formulated in (8.3), can be written as

$$e_{n+1} = \alpha_1 f(P_A e_n) + \alpha_2 \sum_{j=0}^{n-2} P^j f(P_A e_{n-j-1}), \quad n \geqslant 1 \tag{8.23}$$

where

$$\alpha_1 = P_S(0)P_1(S)P_E,$$

$$\alpha_2 = P_2(S)P_E P_1(H)P_S^2(0)$$

and

$$P = P_S(0)P_2(H).$$

Applying Theorem 8.1 with $a_n = \alpha_2 f(P_A e_n)$, and $m = 2$, (so $b_{n+1,m} = e_{n+1} - \alpha_1 f(P_A e_n)$), the increment of e_n can be expressed as

$$e_{n+1} - e_n = \alpha_1 [f(P_A e_n) - f(P_A e_{n-1})] + \alpha_2 f(P_A e_{n-1}) - (1-P)$$
$$\times [e_n - \alpha_1 f(P_A e_{n-1})]$$
$$= \alpha_1 [f(P_A e_n) - f(P_A e_{n-1})] + (1-P)[\alpha_c f(P_A e_{n-1}) - e_n] \tag{8.24}$$

where

$$\alpha_c = \alpha_1 + \alpha_2/(1-P).$$

At this point we introduce the assumption

$$P_2(S) \geqslant P_1(S) \tag{8.25}$$

which will be discussed in Section 5. This entails

$$P_2(S)P_1(H) \geqslant P_1(S)P_2(H)$$

whence

$$\alpha_2 = P_2(S)P_E P_1(H)P_S^2(0) \geqslant P_1(S)P_E P_2(H)P_S^2(0) = \alpha_1 P$$

and

$$\alpha_c = \alpha_1 + \alpha_2/(1-P) \geqslant \alpha_1/(1-P).$$

The equation

$$\alpha_c f(P_A e) = e \tag{8.26}$$

by the concavity of f has at most one nonzero solution. This is finite since f is bounded. Since we assumed $f(e) < e$ for e large, a positive solution exists if and only if the slope of $\alpha_c f(P_A e)$ at 0 is greater than 1, that is, if and only if

$$\alpha_c P_A f'(0) > 1. \tag{8.27}$$

Suppose this holds and denote the solution by e_∞, $0 < e_\infty < \infty$.

Note that if $e_j \leqslant e_\infty$, $1 \leqslant j \leqslant n$, then by (8.23) ($f$ is increasing),

$$e_{n+1} \leqslant \alpha_1 f(P_A e_\infty) + \alpha_2 \sum_{j=0}^{\infty} P^j f(P_A e_\infty) = e_\infty \qquad (8.28)$$

so that if the system starts below e_∞, it remains below e_∞.

To establish convergence to e_∞, we first analyse the increments $\Delta e_n = e_{n+1} - e_n$ of the sequence when the system starts below e_∞. Suppose, in the first case, that the sequence has taken a downward jump (Fig. 8.3a), so $e_n < e_{n-1} < e_\infty$. By the concavity of f, $\alpha_c f(P_A e_{n-1}) - e_n$ is positive, while $\alpha_c[f(P_A e_n) - f(P_A e_{n-1})]$ is negative but smaller in absolute value than $\alpha_c f(P_A e_{n-1}) - e_n$. Since $\alpha_c > \alpha_1/(1 - P)$ (a consequence of (8.25)), $\alpha_1[f(P_A e_{n-1}) - f(P_A e_{n-1})]/(1 - P)$ is also smaller in absolute value than $\alpha_c(P_A e_{n-1}) - e_n$, so from (8.22), $\Delta e_n > 0$, and the next jump of the sequence will be upward.

In the second case, an upward jump has occurred (Figure 8.3b). Usually the jump following this will also be upward but it may happen,

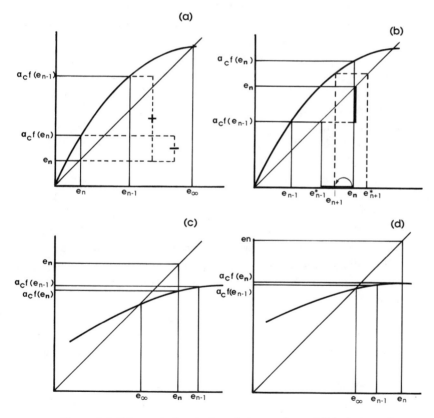

Figure 8.3. Schematic description of changes in established plants for the continuous pasture model.

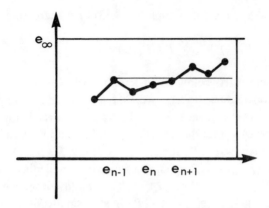

Figure 8.4. A possible trajectory of established plant numbers in the continuous pasture model.

depending on the relative magnitudes of $\alpha_1/(1 - P)$ and α_c, that e_n is relatively much larger than e_{n-1}, and a downward jump can then occur subsequently. In fact (Figure 8.3b) this can only happen if $e_n > e^*_{n-1} \equiv \alpha_c f(e_{n-1})$. If this is so then the next value of the sequence, e_{n+1}, will occur in (e^*_{n-1}, e_n) and hence certainly in (e_{n-1}, e_n). (The shaded lines in Figure 8.3b are of equal length.) What happens next? The downward jump from e_n to e_{n+1} puts the system in Case 1, so the next jump to e_{n+2} will be upward. This may not exceed e_n. If so, it will not exceed e^*_{n+1}, which is greater than e_n. Hence the next jump to e_{n+3} will also be upwards. The system may continue like this, with all subsequent jumps being positive but occurring below e_n. Then the system converges to e, say, $e \leqslant e_\infty$.

On the other hand e_{n+2} may exceed e_n. If it also exceeds e^*_{n+1}, the subsequent jump to e_{n+3} may be negative. But e_{n+3} will be $\geqslant e^*_{n+1}$ and hence $\geqslant e_n$. In any case the whole sequence has advanced past e_n. (Figure 8.4 demonstrates its path in the latter case.) It is clear that the system with this mode of behavior will still converge to a value $e \leqslant e_\infty$.

But taking limits in (8.23), any limit e of e_n satisfies (8.26), to which e_∞ is the unique solution. Thus the system converges to e_∞ if started below e_∞.

The system can be analysed in a similar way when $e_1 \geqslant e_\infty$. In this case $\{e_n\}$ is bounded below by e_∞ and converges to e_∞, usually taking downward jumps (Figures 8.3c, 8.3d). The case when (8.26) has only the zero solution can also be analysed in this way to show that the system converges to zero, that is, the pasture dies out.

Finally, given the behavior of e_n, it is easy to show that seed production x_n and residual hard seed r_n have analogous kinds of behavior. The details are omitted.

(C)   *Cropping models without tillage*
    The case of M cropping years followed by a single pasture year, that is, the case $N = 1$ of (8.14) will be analysed next. We can then write (8.14) in the form

$$e_{n(M+1)+1} = \alpha_3 \sum_{i=0}^{n-1} (P^{M+1})^i f(P_A e_{(n-i-1)(M+1)+1}) + c(P^{M+1})^n$$

$$(8.29)$$

where

$$\alpha_3 = [P_S(0)]^{M+1}[P_2(H)]^M P_2(S) P_E P_1(H),$$

$$P = P_S(0)P_2(H),$$

and

$$c = P_E P_2(S) P_S(0) r_{M+1}.$$

Applying Theorem (8.1) with $a_n = \alpha_3 f(P_A e_{n(M+1)+1})$, $m = 1$, and $b_{n+1,m} = e_{n(M+1)+1} - c(P^{M+1})^n$, gives

$$e_{n(M+1)+1} - e_{(n-1)(M+1)+1} = \alpha_3 f(P_A e_{(n-1)(M+1)+1})$$
$$- (1 - P^{M+1})(e_{(n-1)(M+1)+1} - c(P^{M+1})^{n-1})$$
$$+ c(P^{M+1})^n - c(P^{M+1})^{n-1}$$
$$= \alpha_3 f(P_A e_{(n-1)(M+1)+1}) - (1 - P^{M+1})e_{(n-1)(M+1)+1}$$
$$= [\alpha_M f(P_A e_{(n-1)(M+1)+1}) - e_{(n-1)(M+1)+1}](1 - P^{M+1})$$

where

$$\alpha_M = \alpha_3/(1 - P^{M+1}).\tag{8.30}$$

This system is simple to analyse: if $e_\infty^{(1)}$ denotes the nonzero solution to

$$\alpha_M f(P_A e) = e$$

then clearly the system converges upward to $e_\infty^{(1)}$ if started below $e_\infty^{(1)}$ and downward to $e_\infty^{(1)}$ if started above $e_\infty^{(1)}$. Also $e_\infty^{(1)}$ is positive if and only if

$$\alpha_M P_A f'(0) > 1$$

and if this fails to hold, the system decreases monotonically to zero.
    Now $e_\infty^{(1)}$ is the equilibrium value for established plants in the pasture year following cropping. When $N = 1$, this is the only pasture year. In the general case, $N > 1$, we could let $e_\infty^{(2)} \ldots e_\infty^{(N)}$ be the equilibrium values for established plants in the 2nd, 3rd, ... years following cropping. This system is harder to analyse and we have not worked out the details.

(D)   *More complex systems – without tillage*

As an example of a more complex system to which the methods apply, consider a non-tilled system in which seed production in odd years is 0, and on which another cycle of length 6 is superimposed in which seed production in year $6n + 4$ is C_1 times of that of the standard model amount, production in year $6n + 2$ is C_2 times that of standard model, and production in year $6n$ is zero ($n \geq 0$).

This system can easily be derived from the following more general version of the continuous pasture model, in which we suppose seed production in year n to be $C_n f(P_A e_n)$ rather than $f(P_A e_n)$. From (8.3) we thus have (see below (8.23) for α_1, α_2, and P)

$$e_{n+1} = \alpha_1 C_n f(P_A e_n) + \alpha_2 \sum_{j=0}^{n-2} P^j C_{n-j-1} f(P_A e_{n-j-1}).$$

We recover the above example by taking C_n to be zero except $C_{6n+4} = C_1$ and $C_{6n+2} = C_2$. Thus

$$e_{6n+4} = C_1 \alpha_2 \sum_{j=5,11,\ldots} P^j f(e_{6n+3-j}) + C_2 \sum_{j=3,9,\ldots} P^j f(e_{6n+3-j})$$

$$= C_1 \alpha_2 P^5 \sum_{j=0,1,\ldots} P^{6j} f(e_{6(n-j)-2}) + C_2 \alpha_2 P^3 \sum_{j=0,1,\ldots} P^{6j} f(e_{6(n-j)})$$

and

$$e_{6n} = C_1 \alpha_2 P \sum_{j=0,1,\ldots} P^{6j} f(e_{6(n-j)-2}) + C_2 \alpha_2 P^5 \sum_{j=0,1,\ldots} P^{6j} f(e_{6(n-j-1)}).$$

Solving for e_{6n+4} in terms of e_{6n} gives

$$e_{6n+4} = P^4 e_{6n} + C_2 \alpha_2 P^3 f(e_{6n})$$

which on substitution in the equation for e_{6n} gives

$$e_{6n} = C_1 \alpha_2 P \sum_{j=0,1,\ldots} P^{6j} f[P^4 e_{6(n-j-1)} + C_2 \alpha_2 P^3 f(e_{6(n-j-1)})]$$

$$+ C_2 \alpha_2 P^5 \sum_{j=0,1,\ldots} P^{6j} f(e_{6(n-j-1)}).$$

This is in a form to which Theorem 8.1 applies so

$$e_{6(n+1)} - e_{6n} = \alpha_2 P \{ C_1 f[P^4 e_{6n} + C_2 \alpha_2 P^3 f(e_{6n})]$$

$$+ C_2 P^4 f(e_{6n}) \} - (1 - P^6) e_{6n}.$$

We define a function $g(e)$ by

$$g(e) = \frac{\alpha_2 P}{1 - P^6} \{ C_1 f[P^4 e + C_2 \alpha_2 P^3 f(e)] + C_2 P^4 f(e) \} \qquad (8.31)$$

which, like f, is concave on $e \geq 0$.

As in the previous cropping model the equation

$$g(e) = e$$

has a unique solution which represents the equilibrium point of the system, and the system always converges to this point. The equilibrium point is non zero if $g'(0) > 1$, that is, if

$$\alpha_2 PC_1 f'(0)[P^4 + C_2\alpha_2 P^3 f'(0)] + C_2\alpha_2 P^5 f'(0) - (1 - P^6) > 0.$$

This quadratic expression in $f'(0)$ has one positive root, so there is a nonzero equilibrium point if

$$f'(0) > \frac{(P^6(C_1 - C_2)^2 + 4C_1 C_2)^{1/2}(C_1 + C_2)P^3}{2C_1 C_2 \alpha_2^2}. \tag{8.32}$$

Consider a different but similar system in which all C_n are zero except $C_{6n+2} = C_2$ and $C_{6n+4} = C_1$.

The same analysis leads to the solution of

$$g_2(e) = \frac{\alpha_2 P^3}{1 - P^6}\{C_1 f[P^2 e + C_2\alpha_2 Pf(e)] + C_2 P^2 f(e)\} = e$$

as an equilibrium point. This point is not in general the same as that obtained by solving $g(e) = e$, but it is nonzero if and only if (8.32) holds. Thus both systems persist or die out together, but if they persist they do not in general produce the same amounts of established plants. This curious behavior was observed in computer simulations of cyclical cropping systems and motivated the above explanation.

(E)   *Stochastic versions of the models*

Without attempting to discuss in detail the aims of and problems associated with setting up stochastic population models, or to do justice to the large literature on them, we present here a simple "random environment" version of the model as used in Taylor et al. (1989) and give a couple of its main properties.

We restrict ourselves to the alternate cropping model, that is, the case $N = M = 1$ of (8.14), see (8.8). Assume that seed yield is given in year $2n - 2j - 1$, not by $f(P_A e_{2n-2j-1})$, but by $C_{2n-2j-1} f(P_A e_{2n-2j-1})$, where the C_i are independent and identically distributed positive random variables with mean 1 and finite variance. The model formulated in the previous subsection can be regarded as a special case of this, in which deterministic cyclical oscillations round an average value take the place of the random variation round the mean.

We have from (8.8) (see near (8.23) for α_2 and P)

$$e_{2n+1} = \alpha_2 \sum_{j=0}^{n-1} P^{2j} C_{2n-2j+1} f(P_A e_{2n-2j-1}) + R_n.$$

We can follow exactly the analysis of Section 8.4b to show that each sample has positive increments, assuming the system starts from a fixed point less than e_∞, where e_∞ is the (random) solution to the equation (see (8.35) for α_A)

$$e_\infty = Z_P \alpha_A f(P_A e_\infty) \tag{8.33}$$

and Z_P is the random variable $(1 - P^2) \Sigma_{j \geq 0} P^{2j} C_j$, which has mean 1. (The random series converges because $\Sigma P^{2j} E(C_j)$ does so). Now e_∞ is nonzero for a particular sample if and only if

$$Z_P \alpha_A f'(0) > 1 \tag{8.34}$$

and though this equation is satisfied *on the average*, if $\alpha_A P_A f'(0) > 1$, it need not be for all samples; in those where it fails to hold, the system will die out. For the others, the system converges to e_∞, whose distribution can be obtained from (8.33). In general, unless specified assumptions about the C_i are made, the properties of e_∞ must be obtained by computer simulation. But we note that the average value of the stochastic system will be strictly less than that of its deterministic analogue because of Jensen's inequality (f is concave)

$$E(e_\infty) = E(Z_P) \alpha_A E(f(P_A e_\infty))$$

$$= \alpha_A E(f(P_A e_\infty)) < \alpha_A f(E(P_A e_\infty)).$$

Here we also used the fact that Z_P is independent of $f(P_A e_\infty)$ which follows from the fact that C_{2n+1} is independent of $e_{2n-1}, e_{2n-3}, \cdots$

8.5   Application: the effect of softening rates

The models derived and analysed in this chapter can be applied to questions of agronomic importance regarding regular cropping schemes. There are two basic issues of interest: (i) whether the subclover component of the pasture survives or dies out for the scheme; (ii) the quantity of pasture produced, as measured here by the number of established plants. In general, the effect on survival and production of the pasture of varying the parameters of the model (summer survival, softening and establishment rates, and the form of the density dependence function) and of varying the imposed cropping scheme, require investigation. Here we restrict ourselves to one aspect only, the effect of the changing of softening rates on the pasture production.

We begin by analysing the continuous pasture model. In (8.28) we showed that established plants have a limiting value e_∞ satisfying

$$\alpha_c f(P_A e_\infty) = e_\infty$$

where

$$\alpha_c = P_E P_S(0) P_1(S) + P_E P_S^2(0) P_1(H) P_2(S)/(1 - P_S(0) P_2(H)).$$

As softening rates $P_1(S)$ and $P_2(S)$ vary, α_c will vary, as will the solution e_∞ to (8.26). As a function of $x = P_1(S)$, the softening rate in the first year, α_c has the form

$$\alpha_c(x) = c_1 x + c_2(1 - x)$$

where $c_1 = P_E P_S(0)$ and $c_2 = P_E P_S^2(0) P_2(S)/[1 - P_S(0) P_2(H)]$. Since

$$c_1 - c_2 = P_E P_S(0)[1 - P_S(0)]/[1 - P_S(0) P_2(H)] > 0$$

(we always assume $P_S(0) > 0$), $\alpha_c(x)$ increases in x between the values c_1 and c_2 as shown in Figure 8.5a. The solution e_∞ to (8.26) is positive if $\alpha_c > 1/(P_A f'(0))$, so there are three possibilities: if $c_1 \geqslant 1/(P_A f'(0))$ the pasture survives for any $x \in [0, 1]$, if $1/(P_A f'(0))$ lies in $(c_1, c_2]$, the pasture survives for values of $x \in [x_1, 1]$ (this case is shown in Figure 8.5a), while if $1/(P_A f'(0)) > c_2$ the pasture dies out for any x. When α_c increases, e_∞ increases, as is easily seen from (8.26), and so e_∞ as a function of $x = P_1(S)$ given by

$$\alpha_c(x) f(P_A e_\infty(x)) = e_\infty(x)$$

has the functional form given in Figure 8.5b, again shown for the second case. Thus the long term pasture production can be calculated for given values of the model parameters.

As a function of $x = P_2(S)$, the softening rate in the second year, α_c has the form

$$\alpha_c(x) = c_3 + c_4 x/[1 - P_S(0)(1 - x)]$$

which is also an increasing function of x, so an analogous analysis holds and again e_∞ increases as $P_2(S)$ increases.

Thus: *for the continuous pasture model, long term clover production increases as either (or both) softening rates in the first or subsequent years increase, provided the clover component of the pasture does not die out, when other factors are held constant.*

To demonstrate the joint effect of $P_1(S)$ and $P_2(S)$ on e_∞ in this model, the "standard data" values of the parameters given in Taylor *et al.* (1989) will be used. These are applicable to wheatbelt conditions of Western Australia. A Fortran program was written to solve (8.26) with various values of the parameters, and in Figure (8.6a) we show a contour plot of values of e_∞ obtained when both $P_1(S)$ and $P_2(S)$ vary over the ranges [0, 1], the other parameters being held at their standard values. The values shown are numbers of established plants per square

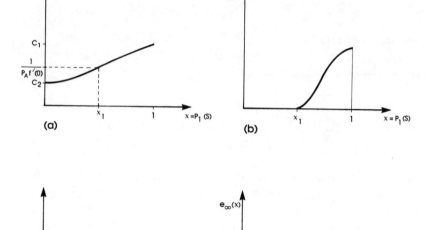

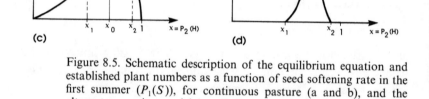

Figure 8.5. Schematic description of the equilibrium equation and established plant numbers as a function of seed softening rate in the first summer $(P_1(S))$, for continuous pasture (a and b), and the alternate cropping model (c and d).

decimeter, and clearly they increase in a regular way with $P_1(S)$ and/or $P_2(S)$.

Next consider the alternating cropping model without tillage, which is the case $N = M = 1$ of (8.14). The relevant equations are

$$\alpha_M f(P_A e_\infty) = e_\infty$$

for the long-term value of established plants, where α_A is given by (8.30) with $M = 1$ as

$$\alpha_A = \frac{P_E P_S^2(0) P_2(H) P_2(S) P_1(H)}{1 - P_S^2(0) P_2^2(H)}. \tag{8.35}$$

As a function of $P_1(S) = 1 - P_1(H)$, this has simple behavior: it decreases as $P_1(S)$ increases. Let us now analyse α_A as a function of $x = P_2(H)$; then

$$\alpha_A(x) = cx(1 - x)/(1 - Px^2)$$

where $P = P_S^2(0)$ and $c = PP_E P_1(H)$. Differentiating, we see that $\alpha_A(x)$ has the form of Figure 8.5c, rising to a maximum value

$$\alpha_A(x_0) = cx_0(1 - x_0)/(1 - Px_0^2) = c/2P = P_E P_1(H)/2$$

at the point $x_0 = [1 - (1 - P)^{1/2}]/P \in (\frac{1}{2}, 1]$ where

$$\alpha_A'(x) = c(Px^2 - 2x + 1)/(1 - Px^2) = 0.$$

Thus there are points $x_1 < x_2 \in (0, 1)$ for which

$$\alpha_A(x_1) = \alpha_A(x_2) = 1/(P_A f'(0))$$

if

$$\alpha_A(x_0) = P_E P_1(H)/2 > 1/(Pf_A'(0)),$$

in which case $e_\infty > 0$, and this case is illustrated in Figure 8.5c. When it occurs, we can see how the solution e_∞ to

$$\alpha_A(x)f(p_A e_\infty(x)) = e_\infty(x)$$

varies as x varies; for $x \in [0, x_1] \cup [x_2, 1]$, there is only the zero solution to this equation, while for $x \in (x_1, x_2)$, $e_\infty(x)$ is positive. Furthermore, for $x \in [x_1, x_0]$, $e_\infty(x)$ decreases with x. Thus it has the form of Figure 8.5b. Our standard data is of this form. In the case when $\alpha_A(x_0) \leqslant 1/(P_A f'(0))$, we have $e_\infty(x) = 0$ for $x \in [0, 1]$ and the pasture dies out. Note that the pasture always dies out for $x = 0$ or $x = 1$, as it must for the alternating system if (i) there is no soft seed to establish plants; or (ii) there is no residual hard seed to regenerate after cropping. Thus: *for the alternating cropping model without tillage, increasing the rate of softening in the first year decreases clover production; whereas increasing the rate of softening in second and subsequent years first increases but then decreases clover production; and there is a unique optimum rate x_0 at which maximum clover production is obtained.* This holds when the other parameters are kept fixed and the pasture does not die out. The contour diagram for varying $P_1(S)$ and $P_2(S)$ together is shown in Figure (8.6b) and has the form expected from the above analysis.

Analyses of more general systems without tillage can also be given. For the case of N pasture years with a single intervening crop year ($M = 1$) and no density dependence, that is, $f(.) = f$, we have from (8.14)

$$e_\infty^{(1)} = \frac{fP_S(0)P_2(S)P_E P_1(H)}{1 - P_S^{N+M}(0)P_2^{N+M}(H)} \frac{[P_S^M(0)P_2^{M-1}(H) - P_S^{N+M}(0)P_2^{N+M-1}(H)]}{[1 - P_S(0)P_2(H)]}.$$

$$(8.36)$$

The contour diagram is shown in Figure 8.6c for the case $N = 3$, $M = 1$, that is, three pasture years followed by one crop year, and is of

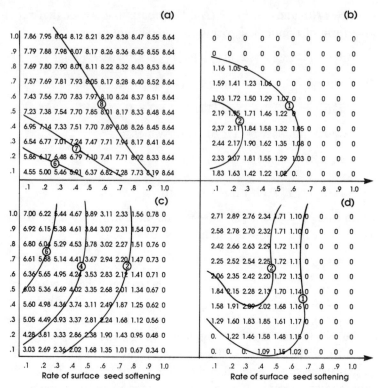

Figure 8.6. Contour plots of established plant numbers for rate of surface seed softening in the first summer (horizontal axis) versus rate of seed softening in second and subsequent years (vertical axis); for continuous pasture (a), alternate cropping model (b), two years pasture and one crop (c), and alternate cropping with tillage (d).

a similar type to the alternate cropping model. But consider the limiting value of established plants in the second year after cropping. From (8.15) we obtain

$$e_\infty^{(2)} = fP_S(0)P_1(S)P_E + e_\infty^{(1)} \tag{8.37}$$

and the contour diagram for $N = 3$, $M = 1$ shows that in this case the plants have almost recovered to their continuous pasture levels. Similarly in the general case the numbers of established plants in years $3, 4, \ldots, N$ after cropping can be written down; they will be even closer to the continuous pasture levels than $e_\infty^{(2)}$.

8.6 Alternate cropping model with tillage

Consider the alternate years cropping model with tillage which was formulated in Section 2. From (8.21) the equilibrium point for established plants in pasture years is given by the solution of

$$e = \frac{bP_E p_1 P_S(D_1)}{1 - a} \left\{ P(S|D_1) + \frac{d}{1 - c} \right\} f(P_A e). \tag{8.38}$$

Comparing the coefficient of $f(P_A e)$ in (8.38) with (8.35), which is the equivalent expression for the alternate cropping system without tillage, we see that the limiting established plants have the same functional form in each case, except that with tillage there is an extra term $d/(1 - c)$ introduced by the latent seed softening, and an extra factor of p_1, which reduces the established plants by the proportion of seed which is distributed by tillage to the first (and only emerging) level. Note that in both cases a is a measure of hardseededness squared, while b is, in both cases, surface hardseededness times surface summer losses, squared.

The contour plot obtained when the probability of softening at the surface and probabilities of latent softening and subsurface seed softening are varied, is shown in Figure 8.6d. This diagram is quite different from the others suggesting that tillage has unexpected effects worthy of further investigation.

BIBLIOGRAPHY

Braakhekke, W. G. (1980). *On Coexistence: a causal approach to diversity and stability in grassland vegetation.* Centre for Agricultural Publishing and Documentation, Wageningen.

Bulmer, M. G. (1984). Delayed germination of seeds: Cohen's model revisited. *Theor. Pop. Biol.* **26** 367–77.

Cohen, D. (1966). Optimizing reproduction in a randomly varying environment. *J. Theoret. Biol.* **12** 119–29.

Cohen, D. (1967). Optimizing reproduction in a randomly varying environment when a correlation may exist between the conditions at the time a choice has to be made and the subsequent outcome. *J. Theoret. Biol.* **16** 1–14.

Cohen, D. (1968). A general model of optimal reproduction in a randomly varying environment. *J. Ecol.* **50** 219–28.

Ellner, S. (1984). Asymptotic behavior of some stochastic difference equation population models. *J. Math. Biol.* **19** 169–200.

Feller, W. (1971). *An Introduction to Probability Theory and its Applications, Vol. 2, 2nd ed.* Wiley, New York.

Gantmacher, F. R. (1959). *The Theory of Matrices, Vol. 1.* Chelsea, New York.

Gates, D. J. and Westcott, M. (1988). Stability of plant mixtures in the absence of infection or predation. *J. Theoret. Biol.* **131** 15–31.

Goh, B-S. (1980). *Management and Analysis of Biological Populations.* Elsevier, Amsterdam.

Guckenheimer, J. and Holmes, P. (1983). *Nonlinear Oscillations, Dynamical Systems and Bifurcations of Vector Fields.* Springer-Verlag, New York.

Hardin, D. P., Taká, P., and Webb, G. F. (1988). Asymptotic properties of a continuous-space discrete-time population model in a random environment. *J. Math. Biol.* **26** 361–74.

Harris, T. E. (1963). *The Theory of Branching Processes.* Springer-Verlag, Berlin.

Hickman, J. C. (1975). Environmental unpredictability and plastic energy allocation strategies in the annual *Polygonum cascadense* (Polygonaceae). *J. Ecol.* **63** 689–701.

Hines, W. G. S. (1987). Evolutionary stable strategies: A review of basic theory. *Theor. Pop. Biol.* **31** 195–272.

Hofbauer, J., Hutson, V., and Jansen, W. (1987). Coexistence for systems governed by difference equations of Lotka–Volterra type. *J. Math. Biol.* **25** 553–70.

Hutson, V. and Moran, W. (1982). Persistence of species obeying difference equations. *J. Math. Biol.* **15** 203–13.

Jiang, H. and Rogers, T. D. (1987). The discrete dynamics of symmetric competition in the plane. *J. Math. Biol.* **25** 573–96.

Jones, F. G. W. and Perry, J. N. (1978). Modelling populations of cyst-nematodes (Nematoda: Heteroderidae). *J. Appl. Ecol.* **15** 349–71.

186

Klemow, K. M. and Raynal, D. J. (1983). Population biology of an annual plant in a temporally variable habitat. *J. Ecol.* **71** 691–703.

Klinkhamer, P. G. L., de Jong, T. J., Metz, J. A. J. and Val, J. (1987). Life history tactics of annual organisms: The joint effects of dispersal and delayed germination. *Theor. Pop. Biol.* **32** 127–56.

Koçak, H. (1986). *Differential and Difference Equations through Computer Experiments.* Springer-Verlag, Berlin.

Kuczma, M. (1968). *Functional Equations in a Single Variable.* PWN-Polish Scientific Publishers, Warsaw.

LaSalle, J. P. (1977). Stability Theory for difference equations. In: Hale, J. ed., *Studies in Ordinary Differential Equations.* M.A.A., Washington, 1–31.

Leon, J. A. (1985). Germination strategies. In: Greenwood, P. J., Harvey, P. H., and Slatkin, M., eds, *Evolution: Essays in honour of John Maynard Smith.* C.U.P., Cambridge, 129–42.

Lessard, S. and Karlin, S. (1982). A criterion for stability–instability of fixation states involving an eigenvalue one with applications in population genetics. *Theor. Pop. Biol.* **22** 108–26.

Levin, S. A., Cohen, D., and Hastings, A. (1984). Dispersal strategies in patchy environments. *Theor. Pop. Biol.* **26** 165–91.

Li, J. (1988). Persistence in discrete age-structured population models. *Bull. Math. Biol.* **50** 351–66.

MacArthur, R. H. (1972). *Geographical Ecology: Patterns in the Distribution of species.* Harper & Row, New York.

MacDonald, N. and Watkinson, A. R. (1981). Models of an annual plant population with a seedbank. *J. Theor. Biol.* **93** 643–53.

Maynard Smith, J. (1982). *Evolution and the Theory of Games.* Cambridge University Press.

Pacala, S. W. (1986a). Neighborhood models of plant population dynamics. 2. Multi-species models of annuals. *Theor. Pop. Biol.* **29** 262–92.

Pacala, S. W. (1986b). Neighborhood models of plant population dynamics. 4. Single-species and multi-species models of annuals with dormant seeds. *Amer. Naturalist.* **128** 859–78.

Pacala, S. W. and Silander, J. A. Jr (1985). Neighborhood models of plant population dynamics. I. Single-species models of annuals. *Amer. Naturalist.* **125** 385–411.

Pollard, J. H. (1973). *Mathematical Models for the Growth of Human Populations.* Cambridge University Press.

Ritland, K. (1983). The joint evolution of seed dormancy and flowering time in annual plants living in variable environments. *Theor. Pop. Biol.* **24** 213–43.

Rossiter, R. C. (1966). The success or failure of strains of *Trifolium subterraneum* L. in a Mediterranean environment. *Aust. J. Agric. Res.* **17** 425–46.

Rossiter, R. C. (1974). The relative success of strains of *Trifolium subterraneum* L. in binary mixtures under field conditions. *Aust. J. Agric. Res.* **25** 757–66.

Rossiter, R. C., Maller, R. A. and Pakes, A. G. (1985). A model of changes in the composition of binary mixtures of subterranean clover strains. *Aust. J. Agric. Res.* **36** 119–43.

Rossiter, R. C. and Palmer, M. J. (1981). An analysis of seed yield in some strains of subterranean clover (*Trifolium subterraneum* L.) when grown in binary mixtures. *Aust. J. Agric. Res.* **32** 445–52.

Sagar, G. R. and Mortimer, A. M. (1976). An approach to the study of the population dynamics of plants, with special reference to weeds. *Appl. Biol.* **1** 1–47.

Schmidt, K. P. and Lawlor, L. R. (1983). Growth rate projection and life history sensitivity for annual plants with a seed bank. *Amer. Naturalist.* **121** 525–39.

Schoener, T. W. (1976). Alternatives to Lotka-Volterra competition: Models of intermediate complexity. *Theor. Pop. Biol.* **10** 309–33.

Schumacher, K. (1983). No-escape regions and oscillations in second order predator–prey recurrences. *J. Math. Biol.* **16** 221–31.

Shapiro, A. P. (1974). A discrete model of the competition of two populations. *Dokl. Biol. Sci.* **28** 406–8.

Szlenk, W. and Zelawski, W. (1985). Plant growth as an iteration process. In *Iteration Theory and its Functional Equations*, Liedl, R., Reich, L., and Targonski, Gy, eds, 183–95. Lecture Notes in Mathematics 1163, Springer-Verlag, Berlin.

Taylor, G. B. (1981). Effect of constant temperature treatments followed by fluctuating temperatures on the softening of hard seeds of *Trifolium subterraneum* L. *Aust. J. Plant Physiol.* **8** 547–58.

Taylor, G. B. (1984). Effect of burial on the softening of hard seeds of subterranean clover. *Aust. J. Agric. Res.* **35** 201–10.

Taylor, G. B., Rossiter, R. C., and Maller, R. A. (1989). A model describing the effect of hard-seededness on the persistence of subterranean clover in wheatbelt leys. In preparation.

Taylor, G. B., Rossiter, R. C., and Palmer, M. J. (1984). Long-term patterns of seed softening and seedling establishment from single seed crops of subterranean clover. *Aust. J. Exp. Agric. Anim. Husb.* **24** 200–2.

Templeton, A. R. and Levin, D. A. (1979). Evolutionary consequences of seed pools. *Amer. Naturalist* **114** 232–49.

Trenbath, B. R. (1976). Models and the interpretation of mixture experiments. In: Wilson, J. R. ed., *Plant Relations in Pastures*. CSIRO, Melbourne, 145–62.

Venable, D. L. and Brown, J. S. (1988). The selective interactions of dispersal, dormancy, and seed size as adaptations for reducing risk in variable environments. *Amer. Naturalist* **131** 360–84.

Venable, D. L. and Lawlor, L. (1980). Delayed germination and dispersal in desert annuals: Escape in time and space. *Oecologia* **46** 272–82.

Watkinson, A. R. (1980). Density-dependence in single-species populations of plants. *J. Theor. Biol.* **83**, 345.

Whitely, D. (1983). Discrete dynamical systems in dimensions one and two. *Bull. London Math. Soc.* **15** 177–217.

de Wit, C. T. (1960). On competition. *Versl. Landbouwk. Onderz.* **66**(8) 1–82.

INDEX

adult plants 49
age dependent softening 2, 106
alleles 46
allocation function 36
annual species 1, 2
   annual plant population 35
asymptotic behavior 25
asymptotically stable (AS) point 29
Australia 1
Bernoulli trials 41
bifurcation 67
binary mixture 1
biomass 37
boundary equilibrium 28
break of season 18
bubble 67, 68
clover, see subterranean clover
coincident isoclines 101
competing strains 3
computer simulations 48, 173
conditions
   for internal equilibrium 34
   for dominance 148
concave 63
configuration 4
contamination 1
constant or average soft seed probabilities
   for model 2, 20
convergence rates 69, 87
   algebraic rate 103, 130
   geometric rate 103, 120, 130
convex 66
coexistence 154
contour diagram 183, 184
crops 158
   alternating pasture-crop system 168
   cropping system 159
   cyclical cropping 160, 178
   wheat 159
   lupins 159
density dependence 4, 40, 159
deWit replacement curve 3
difference equations 15
dispersal pattern 47
domain of attraction 97
dominance 131
   dominance change in Model 1A 148
dominated convergence theorem 75, 116, 117
drought years 39

dynamical system
   dynamical system theory 27
dynamic equilibrium 1
eigenvalue 29, 87
embedding 107
emergent 162
   nonemergent 161
equilibrium point 28
ergodic hypothesis 42
environmental state 38
evolution 37
evolutionarily stable optimum 40
expectation 38, 39
exponential mixtures 12, 132
exponential distribution 48
fecundity 39
flow diagram 6
Fortran program 181
generation 48
genetic structure 46
geometric mean fitness 38
geometric series 22
germination 2, 37, 38, 42
globally asymptotically stable (GAS) point
   29
herbicide 159
high dimensional state space 51
history 6
horizontal asymptote 63
hyperbolic fixed point 97
hyperbolic field function 48
infinite wedge 62
intensity 50
interior equilibrium 28
invariant set 29
invertible 28
isoclines 32
iterate 28, 113
Jacobian matrix 29, 87
Jensen's inequality 180
juveniles 37
key renewal theorem 117, 120
Lagrange stable 31
Lagrange–Sylvester interpolation theorem
   94
latent softening 159, 161
   transition to 169
legume 1
Leslie matrix 47
Liapunov function 30, 86, 98

life cycle 3
life history 3, 37
limit point 119
limiting value 26
linear transformation 62
linearization theorem 29
locally repelling 108
management regime 160
mathematical formulations 11
mean arithmetic fitness 38
medic 2
memory length 20
  finite memory 107, 126
  infinite memory 108, 119
Mediterranean 1
mixtures
  Daliak–Dinninup 132
  Dwalganup–Daliak 132
  Dwalganup–Northam A 132
  Geraldton–Dinninup 132
  Seaton Park–Midland B 132
  Yarloop–Seaton Park 132
model
  applications 133
  bottleneck 44
  comparisons 151
  crossover 43
  deterministic 11
  dispersal 41
  life-history 37
  Model 1 formulation 12–14
  Model 1A formulation 14–16
  Model 2 formulation 16–20
  Model 2A 19–20
  Model G 25, 27, 30–105
  neighborhood 47
  single strain 158
  stochastic 4, 179
  submodel 6
moist years 39
multivalued function 33
neighborhood 48
nematode 45
neutrally stable point 29
nitrogen fixation 158
nonautonomous 26
orbit 28
  orbital drift 59
  orbital point 72
  orbital segment 58
  orbital spiral 59
  unstable orbit 29
outcomes 133, 136, 137
parameters 3
  standard values 131

pasture 1
  alternating 168, 169
  continuous 158
  pasture/crop regimes 160
perennial species 1
Perron–Frobenius theorem 93
persistence 37
phase portrait 138
phenotypic plasticity 39
pigweed 51
plant mortality 39
ploughing 159
Poisson process 48
positively invariant set 29
predator prey interactions 35
preimage 31
probability 4
  conditional 11
  distribution 116
  establishment 11
  softening 11
  summer survival 11, 131
rainfall 39
  high rainfall area 158
random
  environment 37, 179
  proportion 38
  variation 179
  vector 38
rate, *see* probability
ray 56
rectangular hyperbola 62
relative crowding coefficients 6
renewal equation 117
root system 45
rotation 158
seed
  burr seed 2
  dormant seed 2
  free seed 2
  hard seed 2
  i-year-old seed 12, 16
  maximum seed pool 5
  new seed 131
  one-year-old seed 12, 16
  progeny seed 48
  proportion of free seed in Model 2 19
  residual hard seed 2
  seed bank 34
  seed burial 159
  seed output 39
  seed set 47
  seed size 43
  seed softening 131
  seed yields 5

soft seed 2
   total hard seed of all ages 19
   total seed pool 2, 131
   two-year-old seed 12, 16
seedling viability 46
sensitivity analysis 47
single strain 158
single species 51
soil
   fertility 158
   moisture 39
spatial dispersal 43
spectral radius 29
stability concepts 25
stable manifold theorem 97
standard data 181
steady state 29
stochastic 4, 179
strain 1
strong law of large numbers 38
strong persistence 37
subterranean clover 1
survivorship predictor 48

symmetric competition 35
system
   attracted 36
   drawn 36
   permanent 35
   remote 36
target patch 41
Taylor's expansion 74
tillage
   medium tillage 159
toxic compound 1
transient variation 138
transition 10
USA 1
variance 39
velvet leaf 51
vertical asymptote 63
volunteer species 4
weeds 159
weight basis of seeds 7
Western Australia 1

Table A.1. *Index of notation*

Notation	Meaning	Page first used
P_S, Q_S	Probabilities of summer survival, Strains 1, 2	4
$P(S)$, $Q(S)$	Seed softening probabilities, Strains 1, 2	4
P_E, Q_E	Seed estabishment probabilities, Strains 1, 2	4
P_A, Q_A	Probability seedling survives to adulthood, Strains 1, 2	4
P_V	Ratio relative to volunteers	4
M_1, M_2	Monoculture yield of Strains 1, 2	6
k_{12}, k_{21}	deWit constants	6
z	(as proportionate adult plant yield)	6
$P(S\|B)$	Rate of softening of burr seed, Strain 1 (Model 1A)	11
$P(S\|F)$	Rate of softening of free seed, Strain 1 (Model 1A)	11
$P_i(S\|B)$	Rate of softening, i-year-old burr seed, Strain 1 (Models 1, 2)	11
$P_i(S\|F)$	Rate of softening, i-year-old free seed, Strain 1 (Model 1A)	11
x_n, y_n	(as new seed production, Strains 1, 2)	13
r_n, s_n	Residual hard seed in year n, Strains 1, 2	14
t_n, u_n	Maximum seed pool, Strains 1, 2	15
$P_i(F\|B)$	Proportion of i-year-old burr seed freeing, Strain 1 (Model 2)	16
U, V, L, M	Derived parameters for Model G	25
α, β	Derived parameters for Model G	25
X, Y	Derived parameters for Model G	25
p_n, q_n	Normalized seed numbers for Model G	26
ρ	Derived parameter for Model G	27
p_G, q_F	Derived parameters for Model G	33
$\mathscr{R}_p$, $\mathscr{R}_q$	Regions in which trajectories spiral towards axes, Model G	55
R_z	Ray through 0 parameterized by z	56
$\mathscr{C}$	Boundary between $\mathscr{R}_p$ and $\mathscr{R}_q$	58
Δ	Convergence rate for model G	69
a_i, b_i	Normalized softening rates, Model 2	106
X_n, Y_n	Normalised new seed production, Strains 1, 2 (Model 2)	106
$\mathscr{A}$	Invariant attracting curve for Model 2	109
A, B	Derived parameters for Model 2	110
$P_S(0)$	Summer survival probability, single strain model	160
$P_1(S)$	Rate of softening of 1 year-old seed, single strain model	160
$P_2(S)$	Rate of softening of seed more than 1 yo, single strain model	160
$f(x)$	Seed production function, single strain model	160
p_i	Proportion of seed distributed to depth i, single strain model	160
D_i	Depth i, single strain model	161
$P_S(D_i)$	Probability of summer survival at depth i, single strain model	161
$P(S\|D_i)$	Seed softening probability at depth i, single strain model	161

Table A.1. *Continued*

Notation	Meaning	Page first used
$P_i(L\|H)$	Probability hard seed becomes latent, single strain model	162
t_1, t_2	Derived parameters for single strain model	164
a, b, c, d	Derived parameters for single strain model	170
$\alpha_1, \alpha_2, \alpha_c$	Derived parameters for single strain model	174
α_3, α_M	Derived parameters for single strain model	177
α_A	Derived parameter for single strain model	182